SpringerBriefs in Applied Sciences and Technology

For further volumes:
http://www.springer.com/series/8884

Majid Malboubi · Kyle Jiang

Gigaseal Formation in Patch Clamping

With Applications of Nanotechnology

Majid Malboubi
London Centre for Nanotechnology
London
UK

Kyle Jiang
School of Mechanical Engineering
The University of Birmingham
Birmingham
UK

ISSN 2191-530X ISSN 2191-5318 (electronic)
ISBN 978-3-642-39127-9 ISBN 978-3-642-39128-6 (eBook)
DOI 10.1007/978-3-642-39128-6
Springer Heidelberg New York Dordrecht London

Library of Congress Control Number: 2013942150

Printed on acid-free paper

Springer is part of Springer Science+Business Media (www.springer.com)

Preface

This book introduces some novel gigaseal formation approaches in patch clamping using micro-/nanotechnology. Patch clamping is a technique to measure currents passing through ion channels in a cell membrane. It was first introduced by Neher and Sakmann in 1976 and was honored with the Nobel Prize in Physiology or Medicine in 1991. Although this technique has greatly expanded our understanding of the fundamentals of cells, the nature of the technique makes it laborious, time-consuming, and very low in throughput, which does not satisfy the needs of pharmaceutical companies. Patch clamping is totally dependent on the formation of a high-resistance seal, which is as high as giga Ohms and thus known as gigaseal, between cell membrane and patching site. Although the mechanism of gigaseal formation is not yet fully understood, the most important factors in seal formation are known. The introduction of planar patch clamping has revolutionized the technique. The new approach has made use of the advantages of microfabrication processes, microfluidics, and nanotechnology to facilitate patch clamping. Numerous designs have been developed all over the world. While these efforts were successful in developing less laborious and higher throughput systems, the low seal resistances of planar patch clamping systems have so far prevented them from becoming an absolute alternative to the conventional technique. In fact the superior data quality of conventional pipette-based patch clamping recordings has made this approach the gold standard for ion channel studies. It seems that before being able to develop high-throughput systems successful in forming high-resistance seals, in-depth studies on the mechanism of gigaseal formation are needed. In this book, we share with readers our recent practice in acquiring gigaseal formation using micropipettes. Nanotechnology has been used to make this possible.

Contents

Abbreviations

dp/wk	Data Point/Week
DEM	Digital Elevation Model
DMEM	Dulbecco's Modified Eagle's Medium
D_t	Pipette Aperture Size
FBS	Fetal Bovine Serum
FIB	Focused Ion Beam
HEK Cells	Human Embryonic Kidney Cells
HVEM	High Voltage Electron Microscopy
PDMS	Polydimethylsiloxane
PS	Penicillin Streptomycin
RF	Radio Frequency
SEM	Scanning Electron Microscope

Chapter 1
Introduction

A gigaseal is a high-resistance seal in order of giga Ohms formed between cell membrane and patching tool. High-resistance seals are needed in order to be able to record high-quality data from cellular ion channels activities. Gigaseal formation has remained a mystery for decades, as no conditions can warrantee a gigaseal. Although the mechanism of gigaseal formation is not yet fully elucidated, studies have been successful in determining the important conditions that should be met for a gigaseal to form.

In this book the latest micro-/nanotechnology is used to study the influence of factors such as roughness, hydrophilicity and pipette aperture size on gigaseal formation in conventional patch clamping. The major challenges in the study originate from the prerequisites of gigaseal formation, and fall into four categories. First, there are many interrelated factors involved in seal formation. Therefore to study one factor all other factors should be kept unchanged and a large number of experiments are required to ensure that recordings correspond to the factor under study. Second, glass micropipettes used in patch clamping are very fragile. They have a tip size of 1–2 μm and shank of several millimetres. Furthermore the inner wall of a glass micropipette, which interacts with cell membrane, is very difficult to access. All of these confine the number of possible processes that can be used for working on the area of micropipettes, which is involved in seal formation. Third, cleanliness is the most important factor in gigaseal formation and none of the processes that will be used to modify the properties of micropipettes should contaminate their surfaces. Fourth, glass micropipettes are not conductive and it is not possible to coat them with any conductive materials because of the contamination aspect. Therefore focused ion beam milling processes to modify the pipette surface should be done without any coating.

The structure of the following chapters is as follows:

Chapter 2 introduces the conventional patch clamping technique and reviews planar and lateral patch clamping systems. A comparison is made between the three kinds of patch clamping and their limitations and advantages are summarized.

M. Malboubi and K. Jiang, *Gigaseal Formation in Patch Clamping*,
SpringerBriefs in Applied Sciences and Technology,
DOI: 10.1007/978-3-642-39128-6_1,

Chapter 3 reviews the current research progress on gigaseal formation. The mechanism of gigaseal formation and the important factors in seal formation are discussed.

In Chaps. 4, 5 and 6 the effects of roughness, hydrophilicity and pipette aperture size on gigaseal formation are investigated, respectively.

In Chap. 7 glass micropipettes are studied in more detail using various techniques in microscopy and nanotechnology. The effect of pulling parameters on pipette topology and the mechanics of micropipette tip formation are also discussed.

Chapter 2
Development of Patch Clamping

Patch clamping was first introduced into biophysical studies by Neher and Sakmann in [1] 1976 and was soon expanded to many other fields such as biology and medicine. The technique not only allowed the detection of single-channel currents in biological membranes for the first time but also enabled higher current resolution, direct membrane patch potential control, and physical isolation of membrane patches [2]. The development of the patch clamp method was honoured with a Nobel Prize in 1991. The conventional technique of patch clamping uses a glass micropipette to study every single cell individually. The nature of the technique makes it laborious, time-consuming and very low in throughput yet providing high-quality recordings. To achieve higher throughputs mainly two approaches have been adopted over the past decade: automation of conventional patch clamping, and using planar/lateral patch clamp systems. Planar and lateral patch clamping systems take advantage of microfabrication techniques, microfluidics and nanotechnology to overcome many of the difficulties of conventional patch clamping. However the low seal resistances of planar and lateral patch clamping systems have prevented them from becoming a complete alternative to the conventional technique. Because of its superior data quality confirmation of functional activities is often based on the data obtained by conventional patch clamp recordings. However no matter which approach is used, gigaseal formation is the bottleneck in developing high throughput systems capable of producing high-quality recordings.

2.1 Conventional Patch Clamping Technique

In patch clamping a patch of membrane is isolated from the external solution to record the currents flowing into the patch. To achieve this, small glass capillaries are heated and pulled to fabricate glass micropipettes with an aperture size of 1–2 μm. The pipettes are then backfilled with a conductive solution and pressed against the surface of a cell. To improve the sealing condition a gentle suction is

M. Malboubi and K. Jiang, *Gigaseal Formation in Patch Clamping*,
SpringerBriefs in Applied Sciences and Technology,
DOI: 10.1007/978-3-642-39128-6_2, © The Author(s) 2014

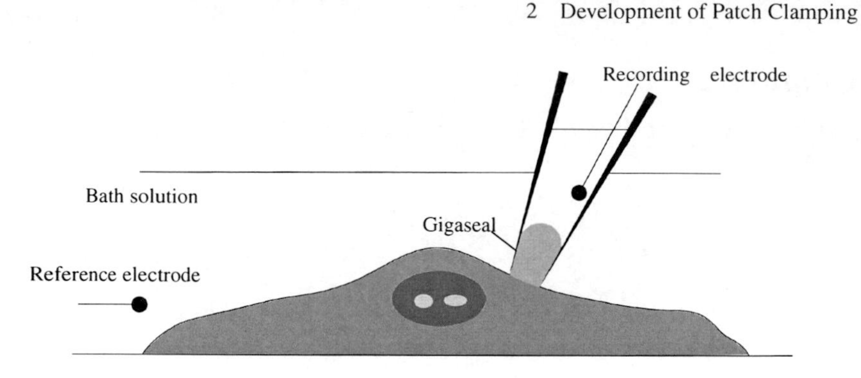

Fig. 2.1 Schematic of the patch clamping experiment. The patch is drawn to have the classical Ω shape; however, recent studies show that the patch can take various shapes [5]

applied to the backend of the pipette. As is shown in Fig. 2.1 there are two electrodes in the patch clamp set-up: a recording electrode inside the pipette and a reference electrode in the bath solution. A high-resistance seal between the glass and the patch of membrane reduces the leakage of current between the two electrodes and completes the electrical isolation of the membrane patch. It also reduces the current noise of the recording, permitting good time resolution of single-channel currents, in the order of 1 pA [3]. Since the electrical resistance of the seal is in the order of giga Ohms, it is called a gigaseal.

2.1.1 Conventional Patch Clamp Configurations

There are several configurations of the patch clamping technique (Fig. 2.2). These configurations enable the technique to study ion channels at different levels, either whole-cell or single ion channels, and to manipulate the fluid on both the extracellular and the intracellular sides of the membrane during experiments [3, 4]. The configurations are:

- Cell-attached mode
- Whole-cell mode
- Inside-out mode
- Outside-out mode
- Perforated patch clamp mode.

2.1.1.1 Cell-Attached Mode

The micropipette is positioned against the cell membrane and suction is applied to the backside of the pipette. A gigaseal forms between glass and cell. This enables the experimenter to study the activities of the ion channels in the patch of

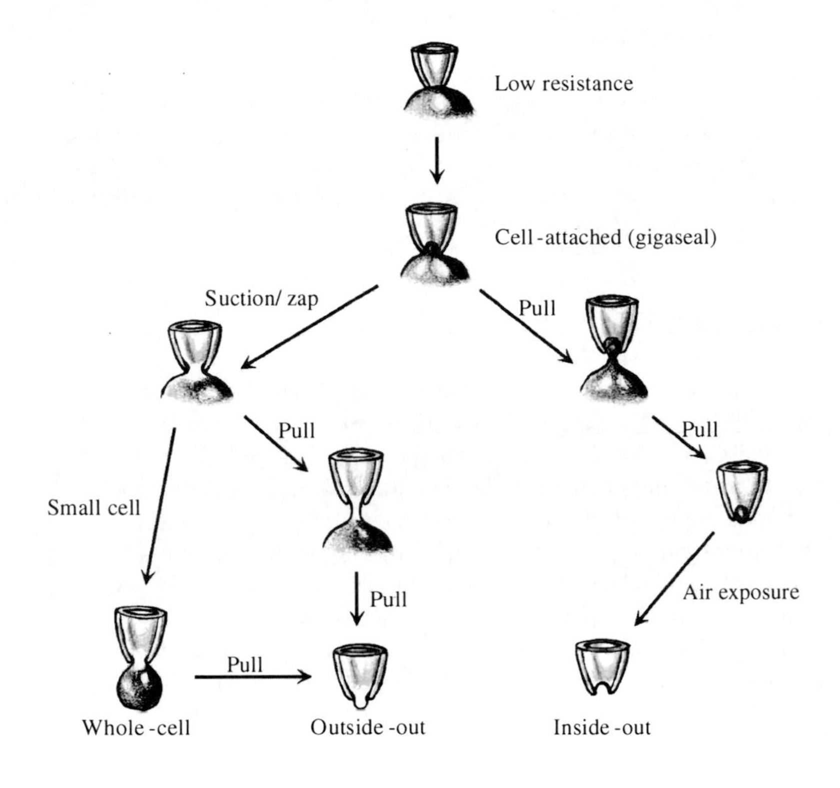

Fig. 2.2 Patch clamp configurations [4]

membrane. The cell-attached patch mode is a single-channel configuration and is the simplest to obtain. Every patch clamp experiment starts with this situation.

2.1.1.2 Whole-Cell Mode

If a higher suction is applied the patch of membrane under the pipette tip in cell-attached mode will rupture and the pipette solution will make direct contact with the cytoplasm. The response of all ion channels within the cell membrane can then be studied.

2.1.1.3 Inside-Out Mode

If the pipette is quickly withdrawn from the cell after the cell-attached mode is obtained, the patch of membrane within the tip of the electrode will tear from the cell while maintaining the gigaseal. This configuration enables study of the effects of intracellular factors on channels. This mode is also a single ion channel configuration.

2.1.1.4 Outside-Out Mode

If the pipette is slowly withdrawn from the cell after whole-cell configuration is obtained, a bleb of cell separates from the cell and forms a patch on the tip of the pipette. This is a single ion channel configuration and because the bath composition can be altered easily during recordings, it enables study of the effects of extracellular factors on the channels.

2.1.1.5 Perforated Patch Clamp Mode

As the volume of the cell is negligible compared with the volume of the patch pipette, the intracellular fluid will be replaced by that of the pipette in the whole-cell configuration. This is a disadvantage when intracellular factors are being studied. To avoid this problem in the cell-attached mode a membrane-perforating agent is added to the pipette solution, which perforates the membrane and allows only small molecules to pass through, so that the cytoplasm's organic composition remains largely intact. This is referred to as perforated patch clamp mode and allows the study of all ion channels in the cell membrane.

There is an equivalent electrical circuit associated with each of these patch clamp configurations. Figure 2.3 shows the electrical components involved in a patch clamping experiment and an equivalent electrical circuit for cell-attached mode.

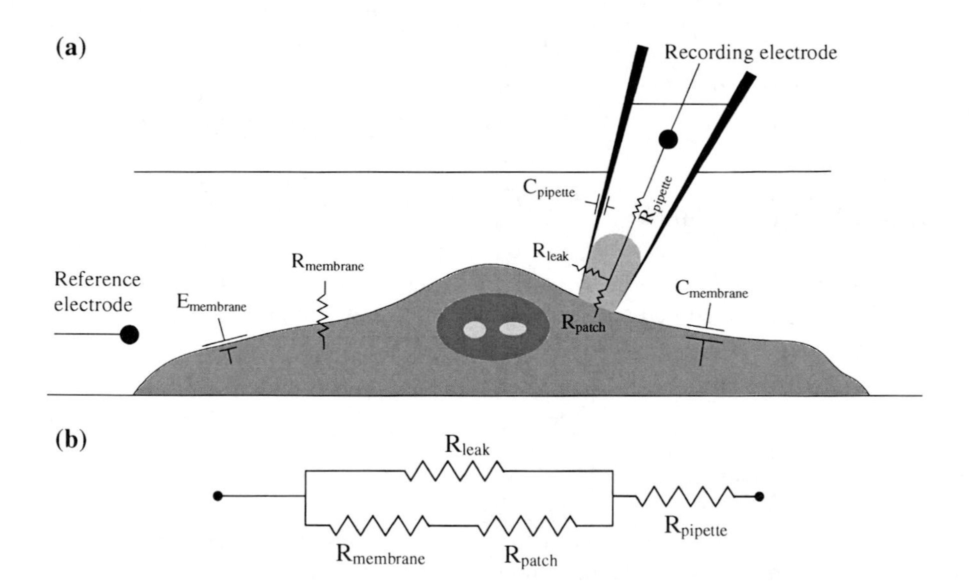

Fig. 2.3 A schematic of patch clamping. **a** electrical components involved in patch clamping, **b** an equivalent circuit for the cell-attached patch configuration

Definitions of the elements in Fig. 2.3 are as below [3]:

$R_{pipette}$ is the pipette resistance.
$C_{pipette}$ is the pipette capacitance.
R_{leak} is the leak resistance and represents the quality of the seal between the glass micropipette and membrane.
R_{patch} is the resistance of the patch of membrane inside the pipette.
R_m is the whole-cell membrane resistance.
C_m is the whole-cell membrane capacitance. A membrane and the intracellular and extracellular media form a capacitor.
E_m is the potential difference across the membrane.

2.2 Other Methods for Measuring Ion Channel Activities

A major constraint in developing new ion channel-based drugs has been the difficulty of screening ion channels at the throughput required of the modern industry in a cost-effective way. The throughput of conventional patch clamping is around 100 data point/week which does not satisfy the needs of pharmaceutical industry [6]. Therefore a good deal of effort has been expended in developing alternative methods of monitoring ion channels activities that can be integrated into industry-standard compound screening formats with corresponding high throughput. The principal methods of high-throughput ion channel screening are: receptor binding assays, flux measurements and fluorescence detection techniques. The advantage of these approaches is their medium to high throughput (15-60 K dp/wk) [7]; however, these techniques measure ion channel activities indirectly and adequate amount of patch clamping experiments are required to validate the results.

2.3 Attempts to Improve Patch Clamping

As was discussed earlier, patch clamping suffers from major drawbacks such as being very laborious and time-consuming, and requiring a lot of experience to get satisfactory results. These drawbacks have made conventional patch clamping a very low-throughput technique. To overcome these problems automated patch clamp systems have been introduced. Automation can reduce the level of complexity and increase the throughput of conventional patch clamping. However, automation of patch clamping presents three major obstacles [8]:

1. An ultraclean surface on a proper substrate is required to be able to obtain gigaseals.
2. A single cell should be positioned on a micron-sized hole without using a microscope and micromanipulator.
3. To automatically perform the complicated steps involved in a patch clamping, complex fluidic and electronic procedures are required.

Efforts to obtain high-throughput patch clamping systems fall into two categories:

- Automation of conventional patch clamping and
- Developing automated planar and lateral patch clamp systems.

In the next two sections, some examples of these attempts are presented. During the past 10 years, many patch clamp systems have been designed. It is not the purpose of the book to review them. A comprehensive review of these efforts can be found in [9].

2.3.1 Automation of Conventional Patch Clamping

A novel idea has been introduced by Lepple-Wienhues et al. [8] for making seals inside a micropipette (Fig. 2.4). The configuration has the potential to be automated and inverts the cell-electrode interface. The pipette is backfilled with a solution containing cells and suction is applied to the tip. The cell forms a seal with the inner side of the pipette. This approach enabled them to develop a fully automated patch clamp robot. Using conventional glass micropipettes for seal formation has some advantages: first, glass is proven to be a good material for seal formation; second, pulling micropipettes from glass tubes is easy and cheap; third, it is a reliable process for making holes at micrometre scale.

Another attempt to improve conventional patch clamping was introduced by D. Vasilyev et al. and is called RoboPatch (Fig. 2.5) [11]. The pipette is inserted into the cell suspension under a positive pressure. A gigaseal is obtained by switching the internal patch pipette pressure from positive to negative (suction). This suction near the vicinity of the patch pipette tip attracts cells to the opening

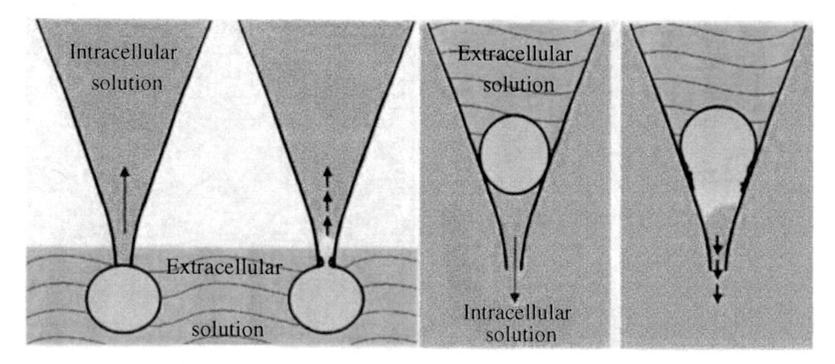

Fig. 2.4 Comparison of conventional and flip the tip patch clamping. Cells are added into glass micropipette and suction is applied to the tip to automatically achieve a gigaseal. Microscope and micromanipulator are no longer required, in contrast to conventional patch clamping [8, 18]

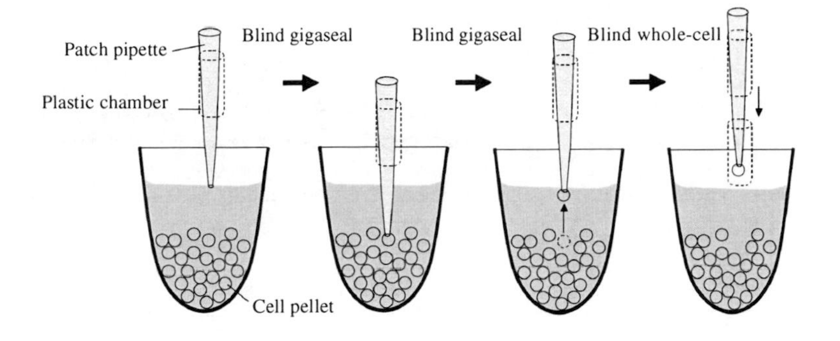

Fig. 2.5 RoboPatch automated system. It utilizes a blind patch voltage-clamp recording method for ion channel drug screening [11]

and eventually results in a physical contact between cellular membrane and patch pipette, allowing the formation of a gigaseal.

Although these approaches overcome some of the drawbacks of conventional patch clamping they still have some limitations. For example, there is no optical access to the cell and the throughput, although higher than conventional patch clamping, does not meet pharmaceutical industry needs. Using a glass micropipette to study an individual cell limits the highest throughput that can be achieved. Applicability of these systems at research level is also limited since they do not have the flexibility often required in research projects.

2.3.2 Planar and Lateral Patch Clamping

In a planar patch clamping process, the pipette is replaced with a micron-sized pore in a flat chip. In lateral patch clamping the pore is in the side wall of the channel. Figure 2.6 compares the three types of patch clamping designs. In both lateral and planar patch clamping, the cell is positioned on the pore and suction is applied to facilitate gigaseal formation. In comparison with conventional techniques, planar and lateral patch clamping configurations do need costly equipment, such as a precise manipulator, a high-magnification microscope and an anti-vibration table. However if the material of the chip is transparent then lateral patch clamping can provide optical access to the patching site.

Fertig et al. [12] conducted one of the earliest attempts in planar patch clamping. Their design is shown in Fig. 2.7. The figure also shows the difference between conventional and planar patch clamp configurations.

The idea of bringing cells to the patching site rather than bringing the pipette to the cells has drawn the attention of many researchers and hundreds of different designs have been developed, each with some advantages over the others. In planar patch clamping the same hole is used for both cell positioning and gigaseal

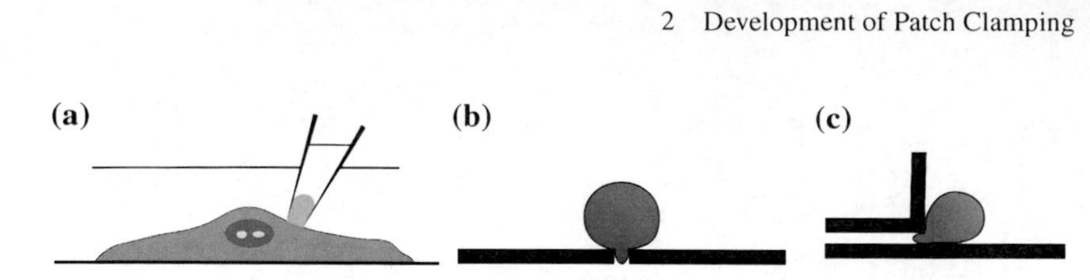

Fig. 2.6 Three types of patch clamping configuration: (**a**) conventional, (**b**) planar and (**c**) lateral

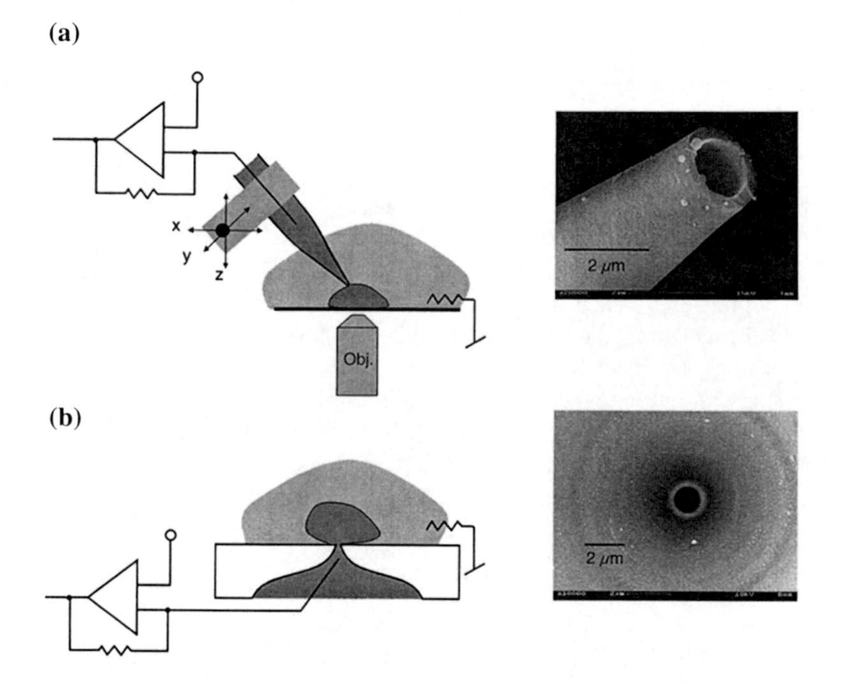

Fig. 2.7 Replacing the patch clamp pipette with a microstructured chip. **a** Whole-cell configuration of the conventional patch clamp technique. The tip of a glass pipette is positioned onto a cell using a micromanipulator and an inverted microscope. **b** Whole-cell recording using a planar chip device having an aperture of micrometre dimensions. Cells in suspension are positioned and sealed onto the aperture by brief suction [12]

formation. This greatly affects the gigaseal quality since debris in the solution may block or contaminate the pore. To overcome this problem, Stett et al. [13] developed a concentric double pipette-like structure (Fig. 2.8). The outer channel is used for cell positioning and the inner channel for current measurements. Positive pressure is initially applied in the inner channel to prevent debris from approaching its surface. Suction in the outer channel directs a cell to the top of the measurement site, which looks exactly like the tip of a pipette. When a cell is placed at the measurement site, suction is used in the inner channel to encourage seal formation.

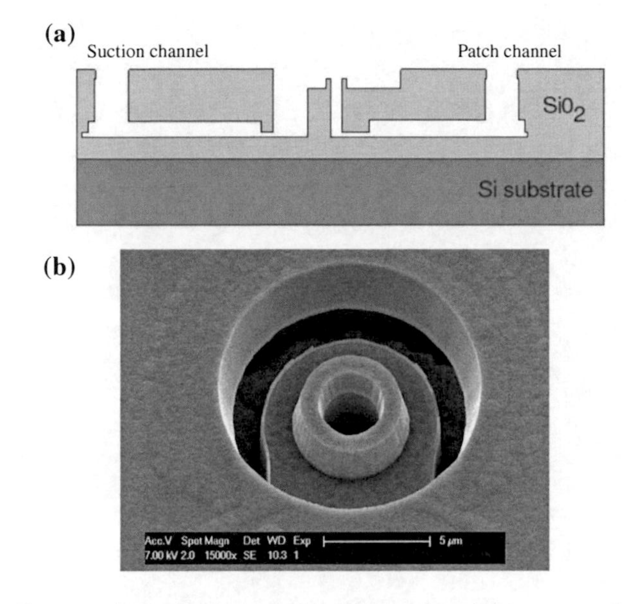

Fig. 2.8 The Cytocentering technology. **a** schematic drawing of the cytocentering site with suction and patch channels; **b** SEM image of two concentric opening formed with focused ion beam milling in a 10 μm thick quartz layer [19]

Planar patch clamping systems increase throughput mainly by taking advantage of their potential to be parallelised and integrated with microfluidic systems. Microfluidic systems facilitate cell manipulation and provide good control on intracellular and extracellular solution exchange. Matthews et al. [14] developed a microfabricated planar patch-clamping system with polydimethylsiloxane (PDMS) microfluidic components (Fig. 2.9).

The design shown in Fig. 2.9 uses the advantages of PDMS as a transparent, easy way to process material. The use of PDMS components has made the exchanging of planar patch clamp chips easier and facilitated quick microfluidic connections. The integration of microfluidics into planar patch clamp systems can enhance the solution control by providing laminar-flow conditions, rapid fluid exchange, and micro-scale control elements such as valves and a pump.

Another advantage of planar patch clamping systems is their potential to be parallelised. Nagarah et al. [15] reported the fabrication of quartz films with high-aspect ratio pores in order to be used as a planar-patch electrophysiology device (Fig. 2.10). The smooth surfaces of pores, the material type and the high depth of pores (which will increase the length of the membrane-glass opposition) resulted in formation of high-resistance seals comparable with those in conventional patch clamping.

Producing planar patch clamp chips is costly and as with patch pipettes, the planar chip cannot be reused once a recording is made. Therefore low-cost and easy fabrication processes are desirable. There is no optical access to the patching cite in planar patch clamping. Lateral patch clamping was introduced to address

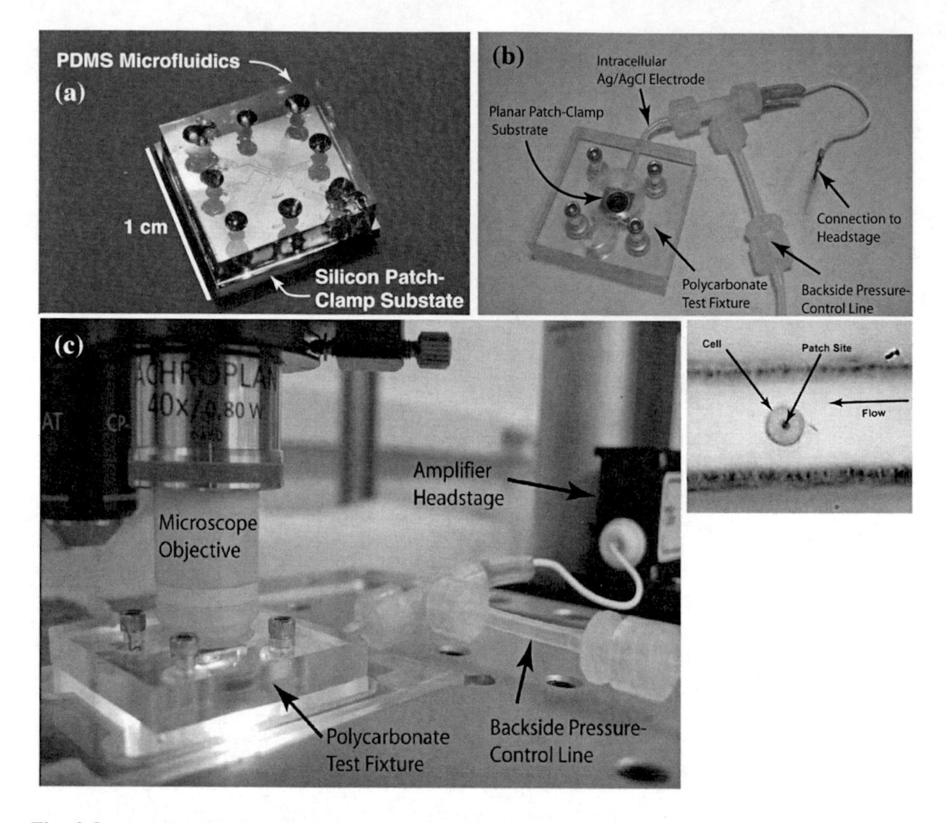

Fig. 2.9 A micro-fabricated planar patch clamp substrate and PDMS microfluidic components. **a** overview of the eight-port microfluidic system, **b** electrodes and capillaries can be simply connectedto the micro-machined planar patch clamp system, **c** use of PDMS has enabled optical access, the inset shows a cell, positioned on the pore. Images were kindly provided by Dr Brian Matthews, Electrical Engineering Department, UCLA

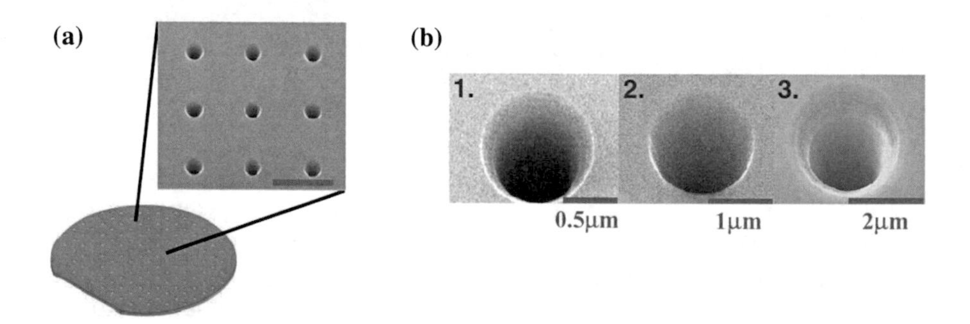

Fig. 2.10 Fabrication of high-aspect ratio pores in quartz films. The pores are polished electrochemically to have smooth surfaces. **a** An SEM image of an array of pores etched 5 μm deep in silicon dioxide. Scale bar = 5 μm. **b** SEM images of high-aspect ratio pores in silicon dioxide with very smooth sidewalls. Pores #1-2 are etched in LTO films while pore #3 is etched in fused quartz. The greater depth of pores increases the length of the membrane-glass interaction in seal formation [15]

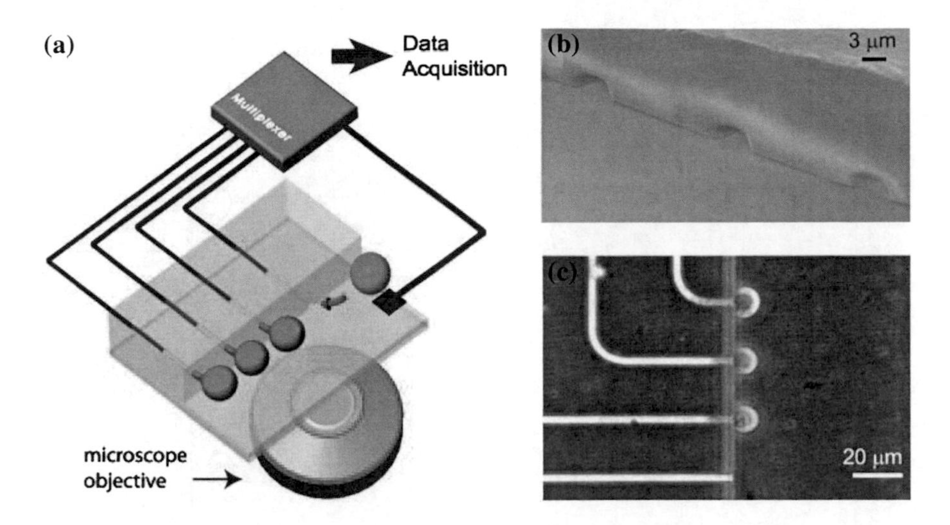

Fig. 2.11 A PDMS-based lateral patch clamping chip developed by Ionescu-Zanetti et al. [16]. **a** Cells were trapped by applying negative pressure to recording capillaries. The device is bonded to a glass coverslip for optical monitoring. **b** Scanning electron micrograph of three recording capillary orifices. The capillary dimensions are 4 × 3 μm. **c** Darkfield optical microscope image of cells trapped at three capillary orifices

some of these issues. PDMS has been widely used in lateral patch clamping systems and offers some unique advantages: firstly, the fabrication process is simple (requiring only moulding and bonding); and secondly, it is transparent, which enables the visualization of the cells during recording.

Figure 2.11 shows a PDMS-based lateral patch clamping system developed by Lee and his colleagues [16].

The device can be easily integrated with microfluidics and provides optical access. Cell-trapping sites in this design are at the bottom plane of the chip which gives the patched cell an unusual deformation. Another design was introduced, by the same group, which has an elevated trapping site [17]. In the new design a 20 μm thick PDMS layer is bonded to the main PDMS microfluidic channels. The success rate of the device is reported to be higher than 80 % but the low values of seals at 250 MΩ are just sufficient for whole-cell recording.

2.4 Comparison of Patch Clamping Methods

In this section a comparison is made between the different kinds of patch clamping. The advantages and drawbacks of each kind are presented in Table 2.1.

As can be understood from the table conventional patch clamping has better data quality and more flexibility but lower throughput, while planar and lateral patch clamping have lower data quality and flexibility and higher throughput.

Table 2.1 Comparison of three kinds of patch clamping: conventional, planar and lateral

	Conventional patch clamping	Planar patch clamping	Lateral patch clamping
Quality of recordings	Superior data quality [11, 15]	Lower data quality [11, 15]	Lower data quality
Throughput	Lower throughput [20–22]	Higher throughput [10, 14, 23, 24]	Higher throughput [16, 17]
Value of gigaseals	Higher [11, 15]	Lower	Lower [16, 17]
Optical access	Yes	No [10]	Yes[a] [16, 25]
Microfluidic integration	N/A	Yes [14, 22, 26–28]	Yes [16, 25]
Ability to control pipette and bath solution	More difficult [29]	Easier [14, 27]	Easier
Density of cell trapping sites	N/A	Lower	Higher [16, 17, 25]
Level of complexity	Higher [17, 21]	Lower [12]	Lower
Level of user expertise	Higher [17, 21]	Lower	Lower
Cost	Expensive equipment, cheap pipettes	Expensive single-use chips [10, 30]	Cheaper than planar, but more expensive than conventional patch clamping
Pipettes/chips fabrication processes	Easy	Difficult	Easier than planar, but more difficult than conventional patch clamping
Time consumption	Higher [12, 17, 21]	Lower	Lower
Ability to choose the best cell for patching	Yes	No	Yes
Potential for automation	Yes [8, 11, 32, 31]	Yes [10, 14, 13, 33, 34]	Yes [16, 17]
Patch clamp configurations	Cell-attached Whole-cell Inside-out Outside-out Perforated	Whole-cell	Whole-cell

[a]If the substrate is made of glass or other transparent materials

References

1. Neher E, Sakmann, B (1976) Single-channel currents recorded from membrane of denervated frog muscle fibres. Nature 260:799–802
2. Hamill OP et al (1981) Improved patch-clamp techniques for high-resolution current recording from cells and cell-free membrane patches. Eur J Physiol 391:85–100
3. Molleman A (2003) Patch clamping: an introductory guide to patch clamp electrophysiology. Wiley, Chichester, 0-471-48685-X
4. Kornreich BG (2007) The patch clamp technique: principles and technical considerations. J Vet Cardiol 9:25–37
5. Suchyna TM, Markin VS, Sachs F (2009) Biophysics and structure of the patch and the gigaseal. Biophys J 97:738–747
6. Owen D, Silverthorne A (2002) Channelling drug discovery current trends in ion channel drug discovery research. Drug Discov World 3:48–61
7. Gill S et al (2003) Flux assays in high throughput screening of ion channels in drug discovery. Assay Drug Dev Technol 1:709–717
8. Lepple-Wienhues A et al (2003) Flip the tip: an automated, high quality, cost-effective patch clamp screen. Recept Channels 9:13–17
9. Dunlop J et al (2008) High-throughput electrophysiology: an emerging paradigm for ion-channel screening and physiology. Nat Rev Drug Discov 7:358–368
10. Klemic K, Klemic J, Sigworth F (2005) An air-molding technique for fabricating PDMS planar patch-clamp Electrodes. Eur J Physiol 449:564–572
11. Vasilyev D et al (2006) A novel method for patch-clamp automation. Eur J Physiol 452:240–247
12. Fertig N, Blick RH, Behrends JC (2002) Whole cell patch clamp recording performed on a planar glass chip. Biophys J 82:3056–3062
13. Stett A et al (2003) Cytocentering: a novel technique enabling automated cell-by-cell patch clamping with the Cytopatchtm chip. Recept Channels 9:59–66
14. Matthews B, Judy JW (2006) Design and fabrication of a micromachined planar patch-clamp substrate with integrated microfluidics for single-cell measurements. J Microelectromech Syst 15 214–222
15. Nagarah JM et al (2010) Batch fabrication of high-performance planar patch-clamp devices in quartz. Adv Mater 22:4622–4627
16. Ionescu-Zanetti C et al (2005) Mammalian electrophysiology on a microfluidic platform. Proc Natl Acad Sci USA (PNAS) 102:9112–9117
17. Lau AY et al (2006) Open-access microfluidic patch-clamp array with raised lateral cell trapping sites. Lab Chip 6:1510–1515
18. Synovo GmbH [Online] Paul-Ehrlich-Str.15 72076 Tübingen Germany. www.synovo.com
19. Cytocentrics Bioscience GmbH. [Online] Joachim-Jungius-Straße 9 18059 Rostock, Germany. www.cytocentrics.com
20. Zhang ZL et al (2008) Fabrication of Si-based planar type patch clamp biosensor using silicon on insulator substrate. Thin Solid Films 516:2813–2815
21. Kusterer J et al (2005) A diamond-on-silicon patch-clamp-system. Diam Relat Mater 14:2139–2142
22. Picollet-D'hahan N et al (ed) (2003) Multi-patch: a chip-based ionchannel assay system for drug screening. In: ICMENS international conference on MEMS, NANO & smart systems, Banff, Alberta Canada, pp 251–254 (2003)
23. Li S, Lin L (2007) A single cell electrophysiological analysis device with embedded electrode. Sens Actuators A 134:20–26
24. Lehnert T, Laine A, Gijs MAM (2003) Surface modification of Sio_2 micro-nozzles for patch-clamp measurements on-chip. In: 7th lnternational conference on miniaturized chemical and blochemlcal analysts systems, Squaw Valley, Callfornla USA, Oct 5–9, 2003, pp 1085–1088
25. Ong W et al (2006) Buried microfluidic channel for integrated patch-clamping assay. Appl Phys Lett 89:093902

26. Shelby JP et al (2003) A microfluidic model for single-cell capillary obstruction by Plasmodium falciparuminfected erythrocytes. Proc Natl Acad Sci 100:14618–14622
27. Sinclair J et al (2003) Stabilization of high-resistance seals in patch-clamp recordings by laminar flow. Anal Chem 75:6718–6722
28. Martinez D et al (2010) High-fidelity patch-clamp recordings from neurons cultured on a polymer microchip. Biomed Microdevices 12:977–985
29. Lapointe JY, Szabo G (1987) A novel holder allowing internal perfusion of patch-clamp pipettes. Eur J Physiol 410:212–216
30. Chen C, Folch A (2006) A high-performance elastomeric patch clamp chip. Lab Chip 6:1338–1345
31. Breguet JM et al (2007) Applications of Piezo-actuated micro-robots in micro-biology and material science. In: Proceedings of the 2007 IEEE international conference on mechatronics and automation. pp 57–62
32. Vasilyev D, Merrill TL, Bowlby MR (2005) Development of a novel automated ion channel recording method using "Inside-Out" whole-cell membranes. J Biomol Screen 10(8):806-813
33. Finkel A et al (2006) Population patch clamp improves data consistency and success rates in the measurment of ionic currents. J Biomol Screen 11:488–496
34. Wilson S et al (2007) Automated patch clamping systems design using novel materials. In: 4M2007: 3rd Internat.Conference on multi-material micro manufacture, Borovets, Bulgaria, 2007

Chapter 3
Gigaseal Formation

To have a good understanding of interactions and forces in gigaseal formation it is necessary to have a close look at the glass and membrane structures.

3.1 Glass Structure

Glass has a molecular structure similar to a quartz crystal. In glass the regular arrangement of atoms found in crystalline quartz is disordered by melting and by the addition of contaminants such as sodium oxide and boric oxide. Quartz is a mineral composed of silicon and oxygen: SiO_2. Each silicon atom in quartz is surrounded by four tetrahedrally disposed oxygens, and each oxygen forms a bridge between two silicon atoms (Fig. 3.1).

Quartz is transparent and strong but with a high melting temperature of 1600 °C. Therefore, it is not suitable for fabricating patch clamping pipettes. Soft glass also known as soda glass is obtained by adding appropriate amount of sodium oxide (Na_2O) to quartz and melts at a lower temperature (800 °C). Sodium ions relatively disorder and loose the glass structure and give a higher conductivity to soda glass as a result soft glass is noisier than quartz or hard glass. Borosilicate glass or hard glass has long been used for fabricating micropipettes and found satisfactory for most purposes. Borosilicate glass has an intermediate structure and properties, between those of fused quartz and soft glass. It is composed of silicon dioxide, sodium oxide and boric oxide (B_2O_3) in relative proportions of about 80, 5 and 15 %. The structure of hard glass is more like that of fused quartz compared with soft glass, as a result it has higher mechanical strength, a higher melting point (1200 °C), lower electrical conductivity and is therefore less noisy [1].

To understand the glass-membrane interactions involved in gigaseal formation surface properties of glass should be considered. The glass surface is composed of silicon atoms and oxygen in one of three configurations: oxygen that forms the bridge between pairs of silicon atoms, oxygen bound to hydrogen and charged oxygen with their charge neutralized by a sodium ion (Fig. 3.2).

M. Malboubi and K. Jiang, *Gigaseal Formation in Patch Clamping*,
SpringerBriefs in Applied Sciences and Technology,
DOI: 10.1007/978-3-642-39128-6_3,

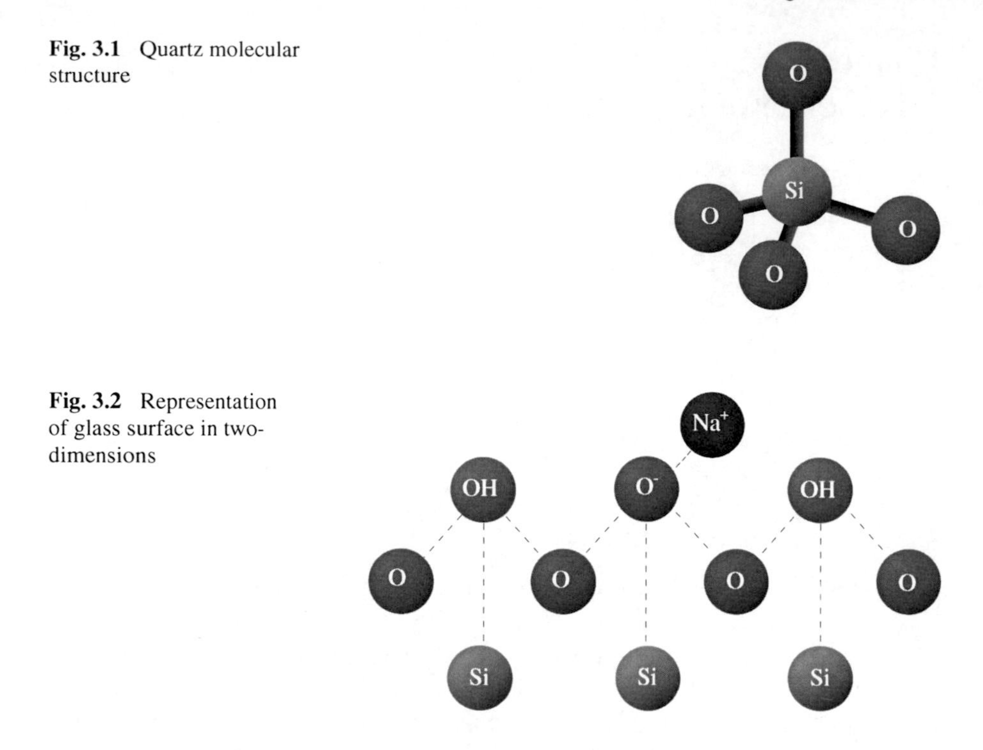

Fig. 3.1 Quartz molecular structure

Fig. 3.2 Representation of glass surface in two-dimensions

Glass has about one oxygen per square nanometre of surface area. In soft glass, about one-third of these oxygen atoms are charged, producing a surface charge density of approximately $0.3/nm^2$; borosilicate glass has fewer charged oxygen atoms and a negative surface charge density of about $0.05/nm^2$. Glass has a negative surface charge and is strongly hydrophilic [1].

3.2 Membrane Structure

Liquid bilayer membranes are constructed of phospholipids, which contain both hydrophobic and hydrophilic residues. In a watery environment phospholipids will arrange themselves spontaneously into structures where the hydrophobic residues face each other [2]. The arrangement found in cell membranes is a bilayer of phospholipids (Fig. 3.3).

The most common phospholipids in animal cell membranes are the phosphoglycerides and sphingomyelin. Together they form about 50 % of the lipid mass in mammalian cells [3]. Sphingomyelin has a free hydroxyl group which can form hydrogen bonds with a molecule of water, the head group of other lipids or with a membrane protein. Phosphoglycerides are present in the bilayer mainly in three forms: phosphatidylethanolamine, phosphatidylcholine and phosphatidylserine. The first two have no net charge, because the positive charge on the alcohol

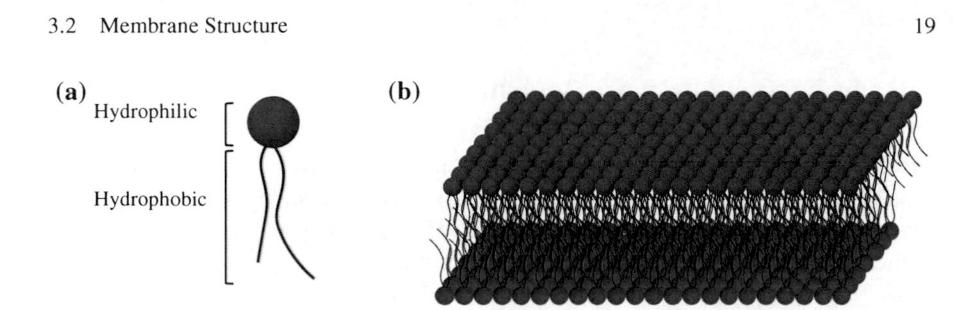

Fig. 3.3 The lipid bilayer. **a** Phosphatidylcholine has a polarized head and fatty tails, **b** Bilayer arrangement of phospholipids in a watery environment

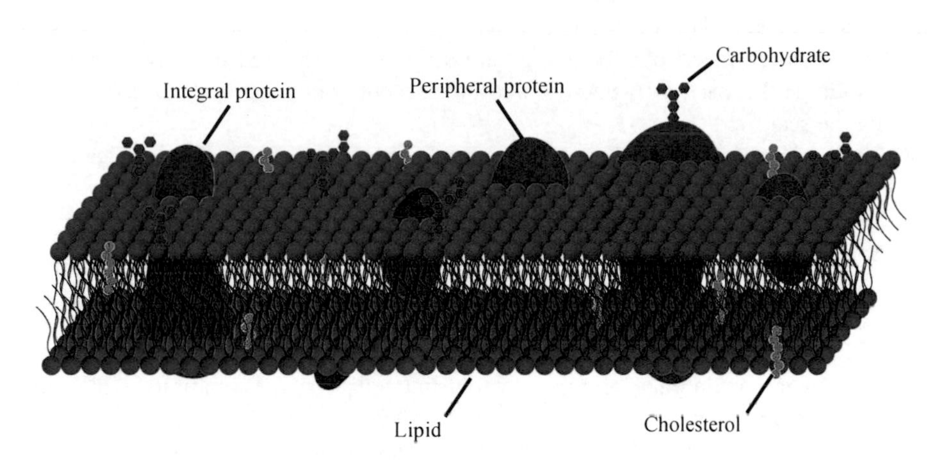

Fig. 3.4 Structure of the cell membrane, showing that it is composed mainly of a lipid bilayer of phospholipid molecules, but with large number of protein molecules protruding through the bilayer

balances the negative charge on the phosphate. The third is negatively charged, therefore the net density of surface charges on phospholipid bilayers is about less than one charge per square nanometre [1].

There are two other general constituents embedded in the lipid bilayer membrane; first, a variable number of membrane proteins whose density ranges from a few hundred per square micrometre up to 10000 per square micrometre and second, a large number of macromolecules associated with the extracellular surface of the membrane (Fig. 3.4).

There are two types of protein in the membrane: integral proteins that protrude all the way through the membrane and peripheral proteins that are attached to only one surface of the membrane and do not penetrate. Many of the integral proteins provide structural channels (or pores) through which water molecules and water-soluble substances, especially ions, can diffuse between the extracellular and intracellular fluid [1, 4].

3.3 Mechanism of Gigaseal Formation

After suction has been applied, a patch of membrane moves down the inside of the pipette [5–8] and a seal forms between cell membrane and glass surface. There have been many studies on the mechanism of gigaseal formation. While earlier studies suggested that seal formation occurs suddenly and there is a direct contact between membrane and glass, more recent studies suggest that the seal is formed gradually and lipid membrane and proteins are mobile in the seal. The main points of agreement and disagreement between studies are listed below:

Agreement:

- Seals form readily between glass electrodes and many types of cell in various physiological and developmental states.
- Gigaseals are mechanically stable, and after seal formation the pipette can be withdrawn to reach to inside-out and outside-out configurations without breaking the seal.
- Positive ions such as Ca^{2+}, Mg^{2+} and H^+ facilitate seal formation significantly.
- Patch pipettes should be clean and used only once.
- The separation between glass and membrane in a gigaseal is in the order of 1Å, i.e. within the distance of chemical bonds.

Disagreement:

- Seal formation is sudden or gradual.
- The integral membrane proteins are detrimental to seal formation or play an important role in it.
- There is direct contact between membrane and glass or there is a thin layer of water between them.
- The patched membrane is detached from or connected to the cytoskeleton.

Gigaseal forms between lipids and glass. The earliest study on seal formation was carried out by Hamill et al. [9]. They demonstrated that a seal forms spontaneously between glass electrodes and different cell types as long as the glass surface and cell surface are clean. Because different cells have quite different extracellular matrix elements and densities and a seal can also form between glass electrodes and pure phospholipid membranes [10], they suggested that a seal forms directly between the glass and the lipid bilayer and seems not to involve integral membrane proteins; in fact, these elements might be detrimental to seal formation.

Membrane patch is initiated by blebbing. In order to explain the mechanisms of seal formation it is important to find out how membrane patches form in the pipette. Because the patch and the pipette tip are at the limits of optical microscope resolution it is rarely possible to see the membrane patch during the experiment [9]. Milton et al. have studied seal formation using pipettes with openings around 10 μm which is almost 10 times the size of pipettes used in patch clamping. In their experiments they were able to reach resistances as high as 100 MΩ and claimed that this is the equivalent of a gigaseal when performing experiments

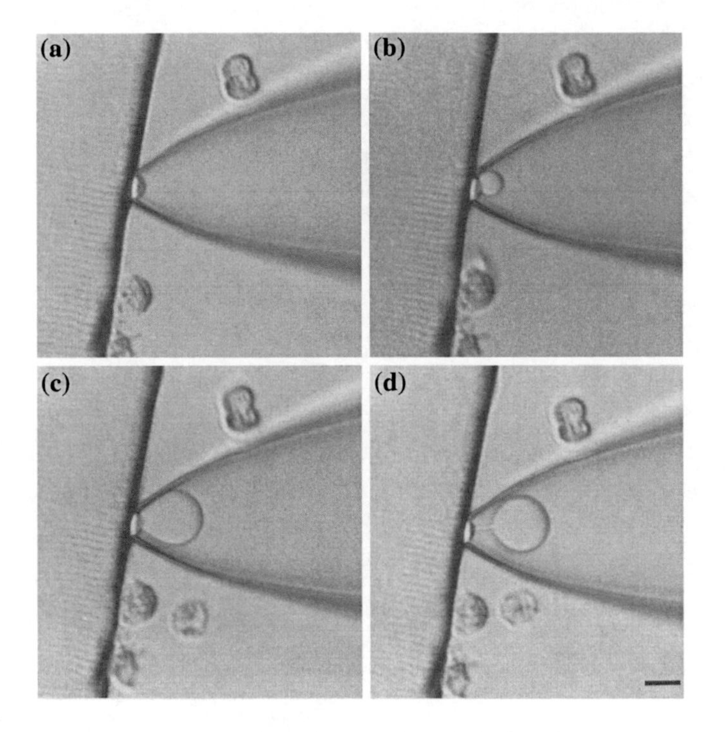

Fig. 3.5 Chronological sequence of bleb formation in a single fibre from the mouse flexor digitorum brevis muscle. **a** The membrane bulged into the pipette after application of suction. **b** Bleb formation at the rim of the pipette. **c** Enlargement of the bleb. **d** Sometimes a thin tether connects the bleb to the cell. The scale bar in d is 10 μm and applies to **a–d** [11]

with normal patch clamping pipettes [11]. Figure 3.5 shows the sequence of patch formation in their experiments.

Milton et al. have suggested that two models can be assumed for patch formation:

1. Native membrane model and
2. Lipid bleb model

These two models are shown in Fig. 3.6.

The native membrane model assumes that the surface membrane is distended and forms a seal with the pipette. The lipid bleb model assumes that when suction is applied tiny bulges occur under the pipette rim and after a critical point is reached phospholipids and highly mobile membrane proteins flow into the region of the bulge and form a bleb. The bleb almost immediately becomes over a micrometre in diameter and would form a gigaseal in a small pipette. The lipid bleb model is in good agreement with the fact that seals form readily between electrodes and many types of cell. This can be explained by the higher lipid content in bleb membranes than a normal surface membrane (with lipid to protein weight ratio of 20:1) and the non-existence of an extracellular matrix in lipid blebs. Because the seal forms between a bleb and a glass surface, single-channel

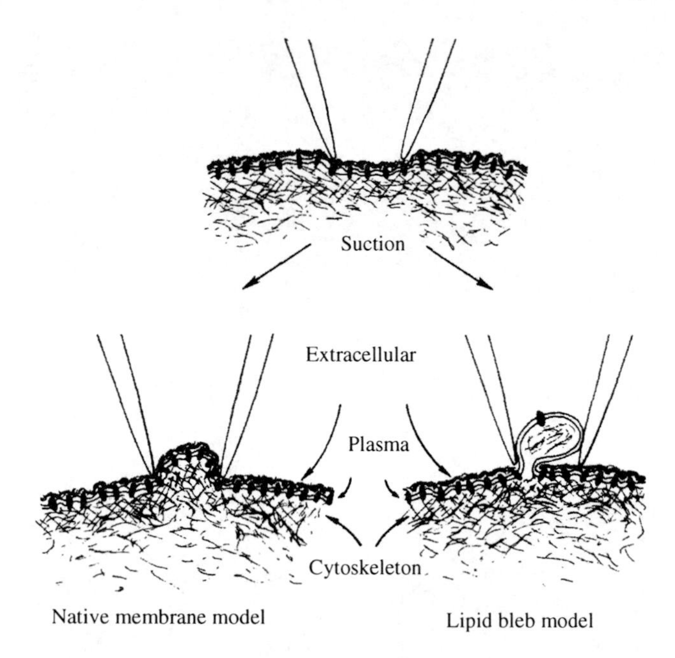

Fig. 3.6 Two models for tight patch formation, native membrane model and lipid bleb model [11]

recordings are from channels in the bleb membrane and may not reflect normal channel behaviour.

The area of the membrane patch increases with suction. This observation is of vital importance in understanding the nature of gigaseal formation. Sokabe et al. demonstrated that the area increase is due to the flow of lipid along the walls of the pipette into the patch [8, 12]. When higher suction was applied the seal region did not move in their experiments, which suggests that attachment of membrane to glass involves membrane proteins. Membrane proteins are denatured against the glass surface, and lipids flow around these anchoring proteins into the pipettes. The gap between the membrane and the glass can be filled with a concentrated solution of the sugar residues of surface glycoproteins. They suggested that because lipid vesicles can also form gigaseals, the seal is a dynamic process that allows sliding. In this model the membrane stretching as a source of area change was ruled out because the area increase is considerably larger than the 2 % elastic limit of lipid membranes. Capacitance measurements confirmed that the area increase in their video images was the result of a net increase in the amount of material comprising the patch. Lorinda et al. have argued that the increase in the patch area is not because of the free flow of lipid but it is the result of the membrane pulling away from the interface of lipid and glass to establish force equilibrium. However both models agree that the seal forms between lipid and glass [10].

Positive ions are required for seal formation. Barkovskaya et al. and Priel et al. found that the presence of positive ions such as Ca^{2+}, Mg^{2+}, Gd^{3+} and H^{+} increases the glass membrane adhesion force and facilitates seal formation [6, 13, 14]. They showed that there is a positive correlation between membrane and glass adhesion force and tight seal formation. No seal was formed when the concentrations of these positive ions were very low. The results of their work show the importance of salt bridges in seal formation. Although these findings can be used practically for immediate seal formation when performing patch clamping experiments, it is not clear whether the increase in the seal is because of higher attraction force per unit area or of a larger contact area.

Gigaseal forms gradually. This suggests that the seal has distributed resistance rather than a local spot weld [12]. In gigaseal formation the membrane patch can moves 5–100 μm down the inside of the pipette. Ruknudin et al. have studied membrane patches inside pipettes using high voltage electron microscopy (HVEM) [7]. In nearly all of their experiments the interior of the pipette walls were covered with membrane. Although they could not measure the distance between the membrane and the pipette because of the thick wall of the pipette, they discovered that the seal is a distributed rather than a discrete structure. Large variations in membrane-pipette length of opposition show that the resistance of the seal is independent of length and the resistance per unit of the seal is very high. Peril et al. suggested that there is a thin layer of water between membrane and glass, which acts as a lubricant and facilitates membrane movement into the pipette. They concluded that a greater length of opposition will result in higher seal values. As mentioned above, the area of contact between membrane and glass may be varied without affecting the gigaseal resistance [13]. The results from Chaps. 4 and 6 of this book also show that the closeness of the membrane to the glass is more important than their length of opposition and that a higher length of opposition does not necessarily result in higher seal values.

The high resistance of the seals suggests that the approach of the glass to the membrane in the area of the seal is in molecular dimensions $\leq$1–2 nm [1, 6]. Considering the structure of the membrane and the glass surfaces, and the small difference of separation, four sources of interaction can be identified [1]:

1. Ionic bonds between positive charges on the membrane and negative charges on the surface.
2. Hydrogen bonds between nitrogen or oxygen atoms in the phospholipids and oxygen atoms on the glass surface.
3. Formation of salt bridges between negatively charged groups on glass and membrane surfaces, which involves interaction of divalent ions such as Ca^{2+}.
4. Van der Waals forces which are due to the close approach of the glass to the bilayer.

The precise order of importance of each of these four interactions is not known [1].

Suchyna et al. demonstrated that there is a 1–2 μm region of the seal below the patch dome where membrane proteins are excluded. This region forms the gigaseal (Fig. 3.7) [15].

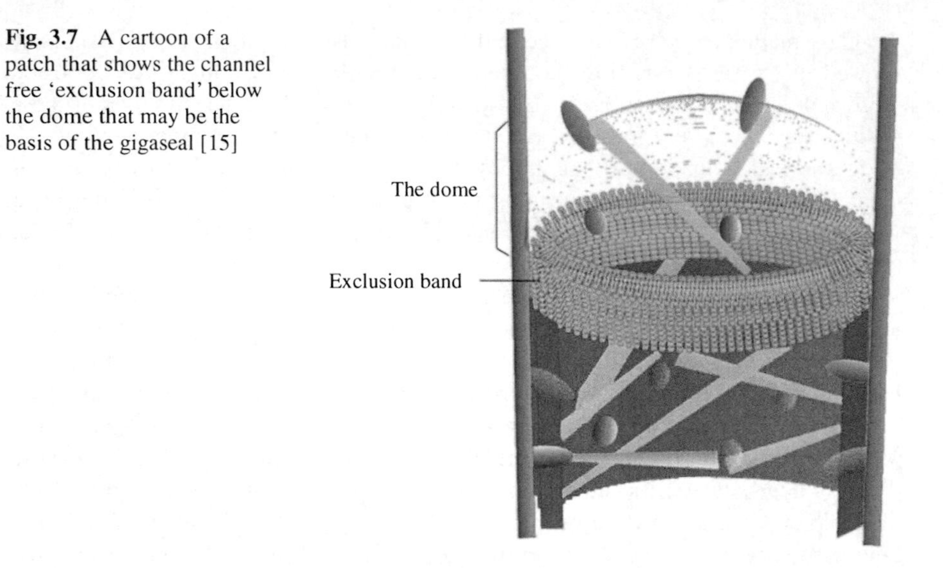

Fig. 3.7 A cartoon of a patch that shows the channel free 'exclusion band' below the dome that may be the basis of the gigaseal [15]

Their model is consistent with most of the research described earlier:

1. The seal occurs between pure lipid bilayer and glass, which explains reports on seal formation between different cell types and pure lipid vesicles. The primary attractive force for the formation of the gigaseal is van der Waals attraction which is overlaid with electrostatic repulsion. Since the van der Waals forces are not chemically specific this may explain the seal formation between glass and different cell types [9], lipid bilayers [10] and rubber [14].
2. The patch formation starts with blebbing. Upon applying suction, membrane blebs into the pipette and forms a patch.
3. The patch is connected to the cytoskeleton. Within seconds to minutes the cortex reforms beneath the dome and the cytoskeleton adapts to new stresses and to the physical constraints of the tip.
4. The seal is distributed over the pure lipid bilayer region. As was discussed earlier, the resistance pure unit length of area of membrane-glass contact is very high.
5. Some studies have claimed that the close opposition of lipid bilayer and glass cannot be obtained in the presence of integral proteins. In this model membrane proteins protrude above the bilayer by a few nanometres. These proteins are denatured against the glass surface and pull the membrane closer to the glass (Fig. 3.8). Therefore this model shows that a gigaseal can be obtained while the proteins are present in the seal.
6. Because of the denaturation process of membrane proteins the seal happens gradually.
7. The pure lipid bilayer region below the dome suggests that the seal is distributed. Channels in the patch can diffuse from the seal to the dome or vice versa.

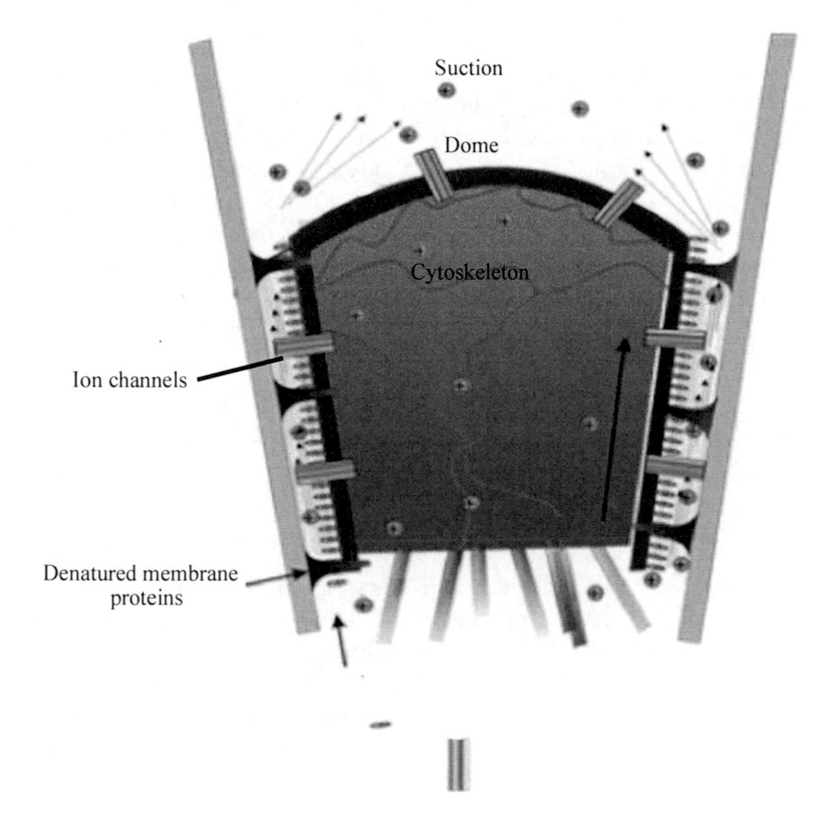

Fig. 3.8 Cartoon of patch structure. There are three regions in the patch: the dome, the gigaseal between the membrane and the glass, and the cytoskeleton. Proteins sticking far from the bilayer are denatured against the glass, pulling the membrane closer to the glass. Ion channels are distributed in varying density throughout the dome and the seal. Ion channels in the patch can diffuse from the seal to the dome or vice versa [15]

3.4 Important Factors in Gigaseal Formation

A large number of parameters affect gigaseal formation, making it difficult to study seal formation. This section gives a summary of the most important factors, while some unknown factors may still cause the seal to form randomly.

3.4.1 Cleanliness

Cleanliness of glass micropipette and plasma membrane is the most important factor in gigaseal formation, as has been emphasized in the literature [9, 16, 17]. The crux of a successful seal is that the cell membrane is reached without damage to, or contamination of, the pipette tip and that the contact with the membrane is full and even; therefore there are many considerations involved in protecting the pipette from contamination. It is well known that once a cell (or debris in the solution) seals to the tip of a glass pipette or to the aperture of a planar patch clamp device, a residue, which is very difficult to remove, will stay at the hole and prevent the subsequent formation of another gigaseal. Therefore positive pressure on the pipette fluid is required to keep the tip from contamination by debris in the bath. In conventional patch clamping before lowering the pipette into the bathing solution, there must be a slight pressure on the pipette fluid to blow any contaminations in the bathing solutions away from the pipette tip. These contaminations often gather in the bathing solution at the fluid-air interface, so pressure must be on before this is crossed [18]. The patch pipette should be made and used immediately to reduce tip contamination and subsequent bad sealing properties.

3.4.2 Roughness

Roughness is a very important factor in gigaseal formation [19–24]. A rough, sharp pipette can easily destroy the cell. Fire polishing is used usually to produce a soft and smooth tip [25, 26]. However, fire polishing also increases the probability of contamination because of its blunting effect on the tip. In planar patch clamping, processes used to fabricate chips were carefully selected or modified to produce a smooth patching site [17, 19–21, 27]. The effect of roughness on gigaseal formation is the subject of Chap. 4.

3.4.3 Hydrophilicity

Hydrophilicity of the patching site is another important issue in seal formation [28]. The hydrophilic cell membrane will not spontaneously interact with the hydrophobic surface in a way to form gigaseals. Different treatment methods were used in the literature to increase the hydrophilicity of the patching site. The effect of hydrophilicity on gigaseal formation is the subject of Chap. 5.

3.4.4 Tip Size

It is common knowledge that in practice pipettes with a smaller opening form a better seal and lower leakage current. In planar patch clamping tip size has been decreased to increase the seal resistance [29–32]. It should however be noticed that decreasing the tip size will also decrease the chance of having ion channels in the patch, and it is also more difficult to rupture the membrane to obtain whole-cell configuration. The effect of tip size on gigaseal formation is discussed in Chap. 6.

3.4.5 Roundness

The patching site should have a circular aperture without sharp corners to allow the cell to cover the pore evenly [17, 19–21, 33]. Lau et al. show that higher resistances can be achieved by more rounded apertures [33]. Roundness of glass micropipettes is measured using the nanotomography technique and the results are presented in Chap. 7.

3.4.6 Other Factors

There are many other factors that are important in gigaseal formation, such as: the biological condition of the cell, the cleanliness of the surface of the cell, pH and presence of divalent ions, the experience and patience of the operator, mechanical vibration, the angle of approach of the pipette to cell, etc.

References

1. Sakmann B, Neher E (2009) Single-channel recording, 2nd edn. Springer, Vienna 978-1-4419-1230-5
2. Lodish H et al (2007) Molecular cell biology. W.H. Freeman & Co Ltd, New York
3. Alberts B et al (2008) Molecular biology of the cell. Garland Science, New York
4. Guyton AC, Hall JE (2000) Text book of medical physiology. W.B. Saunders company, London, 0-7216-8677-X
5. Nagarah JM et al (2010) Batch fabrication of high-performance planar patch-clamp devices in quartz. Adv Mat 22:4622–4627
6. Priel A et al (2007) Ionic requirements for membrane-glass adhesion and gigaseal formation in patch-clamp recording. Biophys J 92:3893–3900
7. Ruknudin A, Song MJ, Sachs E (1991) The ultrastructure of patch-clamped membranes: a study using high voltage electron microscopy. J Cell Biol 112:125–134
8. Sokabe M, Sachs F, Jing Z (1991) Quantitative video microscopy of patch clamped membranes stress, strain, capacitance, and stretch channel activation. Biophys J 59:722–728
9. Hamill OP et al (1981) Improved patch-clamp techniques for high-resolution current recording from cells and cell-free membrane patches. Eur J Physiol 391:85–100
10. Opsahl LR, Webb WW (1994) Lipid-glass adhesion in giga-sealed patch-clamped membranes. Biophys J 66(1):75–79
11. Milton RL, Caldwell JH (1990) How do patch clamp seals form? A lipid bleb model. Eur J Philos 416:758–765
12. Sokabe M, Sachs E (1990) The structure and dynamics of patch-clamped membranes: a study using differential interference contrast light microscopy. J Cell Biol 111:599–606
13. Barkovskaya DAY et al (2004) Gadolinium effects on gigaseal formation and the adhesive properties of a fungal amoeboid cell, the slime mutant of neurospora crassa. Membr Biol 198:77–87
14. Sachs F, Qin F (1993) Gated, ion-selective channels observed with patch pipettes in the absence of membranes: novel properties of a gigaseal. Biophys J 65:1101–1107
15. Suchyna TM, Markin VS, Sachs F (2009) Biophysics and structure of the patch and the gigaseal. Biophys J 97:738–747

16. Kornreich BG (2007) The patch clamp technique: Principles and technical considerations. J Vet Cardiol 9:25–37
17. Stett A et al (2003) CYTOCENTERING: A novel technique enabling automated cell-by-cell patch clamping with the CYTOPATCHTM chip. Receptors Channels 9:59–66
18. Molleman A (2003) Patch clamping: an introductory guide to patch clamp electrophysiology. Wiley, Chichester, 0-471-48685-X
19. Fertig N, Blick RH, Behrends JC (2002) Whole cell patch clamp recording performed on a planar glass chip. Biophys J 82:3056–3062
20. Li S, Lin L (2007) A single cell electrophysiological analysis device with embedded electrode. Sens Actuators A 134:20–26
21. Ong WL, Yobas L, Ong WY (2006) A missing factor in chip-based patch clamp assay: gigaseal. J Phys 34:187–191
22. Malboubi M et al (2009) The effect of pipette tip roughness on giga-seal formation. World Congr Eng 2:1849–1852
23. Malboubi M et al (2009) Effects of the surface morphology of pipette tip on Giga-seal formation. Eng Lett 17:281–285
24. Malboubi M, Gu Y, Jiang K (2011) Surface properties of glass micropipettes and their effect on biological studies. Nanoscale Res Lett 6:1–10
25. Goodman M, Lockery SR (2000) Pressure polishing: a method for re-shaping patch pipettes during fire polishing. J Neurosci Methods 100:13–15
26. Yaul M, Bhatti R, Lawrence S (2008) Evaluating the process of polishing borosilicate glass capillaries used for fabrication of in vitro fertilization (iVF) micro-pipettes. Biomed Microdevices 10:123–128
27. Matthews B, Judy JW (2006) Design and fabrication of a micromachined planar patch-clamp substrate with integrated microfluidics for single-cell measurements. J Microelectromech Syst 15:214
28. Alberta C (2003) Multi-patch: a chip-based ionchannel assay system for drug screening. In: Picollet-D'hahan N et al (ed.) ICMENS international conference on MEMS, NANO and smart systems, pp 251–254
29. Klemic K, Klemic J, Sigworth F (2005) An air-molding technique for fabricating PDMS planar patch-clamp Electrodes. Eur J Physiol 449:564–572
30. Kusterer J et al (2005) A diamond-on-silicon patch-clamp-system. Diamond Relat Mater 14:2139–2142
31. Petrov AG (2001) Flexoelectricity of model and living membranes. Biochimica et Biophysica Acta 1561:1–25
32. Petrov AG (2006) Electricity and mechanics of biomembrane systems: flexoelectricity in living membranes. Analytica Chimica Acta 568:70–83
33. Lau AY et al (2006) Open-access microfluidic patch-clamp array with raised lateral cell trapping sites. Lab Chip 6:1510–1515

Chapter 4
Effect of Roughness on Gigaseal Formation

Surface roughness is one of the most important factors in gigaseal formation and its effect has been emphasized in the literature [1–8]. A rough pipette tip in conventional patch clamping, or patching site in planar patch clamping prevents seal formation. In this chapter the effect of roughness on gigaseal formation is discussed.

4.1 Glass Micropipette Fabrication

Consistent terminology is required when one discusses micropipettes. To avoid confusion, some terms used in this book relating to micropipette terminology are described here (see Fig. 4.1):

Tip is the very end of the pipette.
Tip size or *aperture size* is the inner diameter of the pipette tip.
Shank is the tapered segment of the pipette.
Shaft is the straight portion of the capillary tubing.

Glass micropipettes are fabricated through a heating and pulling process using a puller machine. The glass tube is heated and pulled while it is softened. The process is repeated in several stages until the tube is pulled apart. The connection is then broken by a final hard pull. To have better control of the process, the last pull takes place while the glass is not being heated and therefore it is called hard pull [9]. The puller used in the experiments is a flaming/brown micropipette puller machine (Model P-97, Sutter Instruments, Novato, CA) (Fig. 4.2). The filament of the puller machine was FB230B (2.0 mm square box filament, 3.0 mm wide, Sutter Instruments).

A typical PULL CYCLE is described below (see Fig. 4.3):

- The heat is turned on.
- The glass heats up and a weak pull draws the glass out until it reaches the programmed velocity.

M. Malboubi and K. Jiang, *Gigaseal Formation in Patch Clamping*,
SpringerBriefs in Applied Sciences and Technology,
DOI: 10.1007/978-3-642-39128-6_4, © The Author(s) 2014

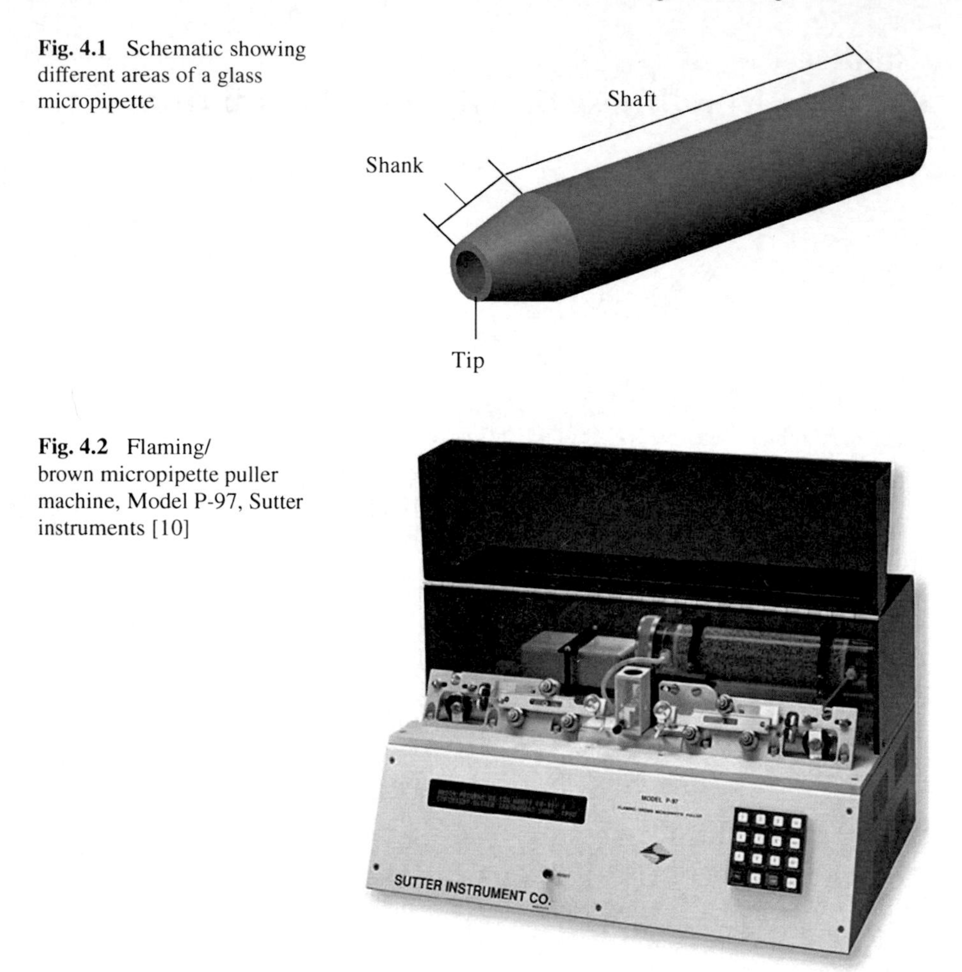

Fig. 4.1 Schematic showing different areas of a glass micropipette

Fig. 4.2 Flaming/brown micropipette puller machine, Model P-97, Sutter instruments [10]

- When the programmed velocity has been reached, the heat is turned off and the air is turned on.
- If DELAY is >0 the air is activated for 300 ms and the hard pull is activated after the specified DELAY.

4.2 Measurement of Roughness

Hard pull results in a sharp tip with jagged edges. High-magnification scanning electron microscope (SEM) images revealed the surface nature of the pipette tip to be in contact with cells (Fig. 4.4). The first step in investigating the effect of roughness on gigaseal formation is to measure the pipette's roughness. This measuring will give a clue about the value of the roughness and how comparable it is with the sizes of the cell and its components.

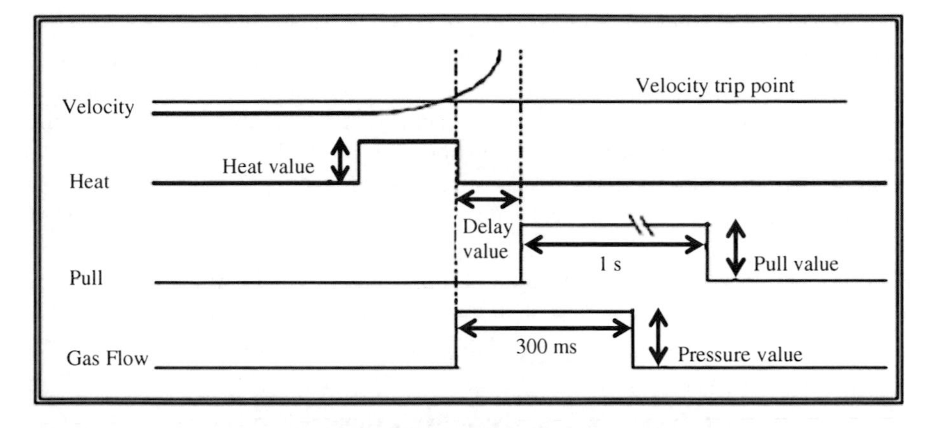

Fig. 4.3 A typical pull cycle in a programme of a puller machine [9]. There are two cooling parameters: delay and time. This illustration represents the pulling cycle when delay is active. For details on glass micropipette pulling and on the definition of each parameter see Sect. 7.2.1

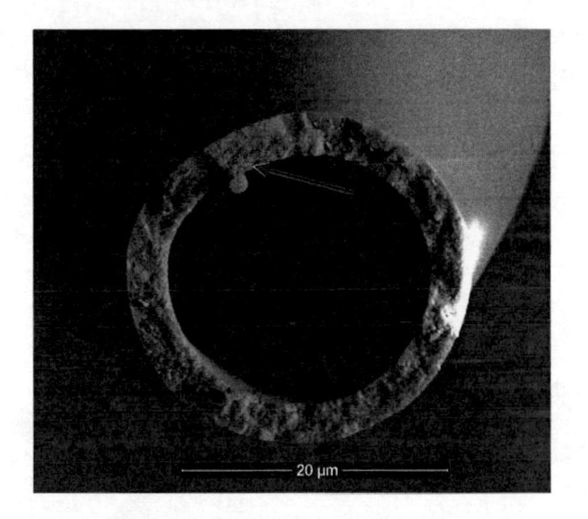

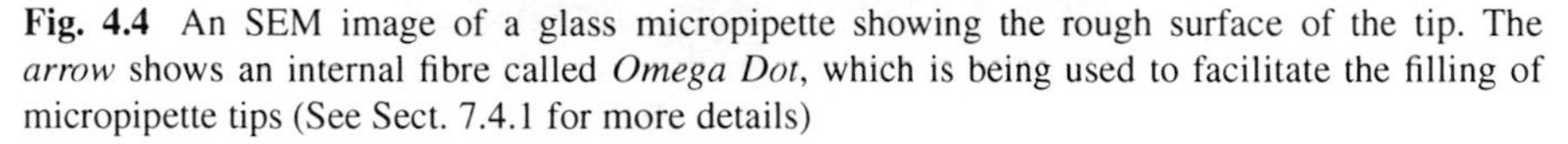

Fig. 4.4 An SEM image of a glass micropipette showing the rough surface of the tip. The *arrow* shows an internal fibre called *Omega Dot*, which is being used to facilitate the filling of micropipette tips (See Sect. 7.4.1 for more details)

One of the challenges of working with glass micropipettes is their fragility. For different applications, from the patch clamping to microinjection, they may have an aperture size of anything from a few hundreds of nanometres to a couple of micrometres and a shank of several millimetres to a couple of centimetres. This makes micropipettes very fragile and less manoeuvrable. Special care should therefore be taken when working with micropipettes.

4.2.1 SEM Stereoscopic Technique

The three-dimensional structure of a specimen is projected into a two-dimensional plane during the process of scanning in electron microscopes. To determine the three-dimensional structure of pipette tips, the SEM stereoscopic technique [11–13] is used. The stereoscopic technique scans the same area of the object from different angles by tilting the object with respect to the electron beam. As a result surface features at different heights have different lateral displacements. By measuring the parallax movement of features from their location in the first image to their new location in the second image, depth can be calculated [14].

In practice capturing SEM stereo images can be challenging. The quality of images is very important in the calculation of accurate 3D data. Many factors should be taken into account when attempting to capture SEM images which satisfy stereoscopic technique requirements, such as [15]:

- Illumination: capture images under optimal illumination. The image should not be too dark or too bright.
- Sharpness: images should be captured with maximum sharpness.
- Disparity: The software package used here (MeX) calculates the depth image based on the disparity in the stereo image. Disparity can be increased by using higher magnifications and/or higher tilting angles.
- Voltage: the voltage of the electron beam should be well chosen. If the chosen electron beam voltage is too high then it is possible for the electron beam to intrude into the surface of the object, so that details which lie inside the object are also received. This circumstance would lead to a wrong reconstruction of the surface.

To capture high-quality SEM images which satisfy stereoscopic technique requirements, glass micropipettes were coated with a thin layer of platinum (<5 nm). The machine used for capturing SEM images was an FEI dual-beam focused ion beam system (FEI, Hillsboro, Oregon). Figure 4.5 shows the position

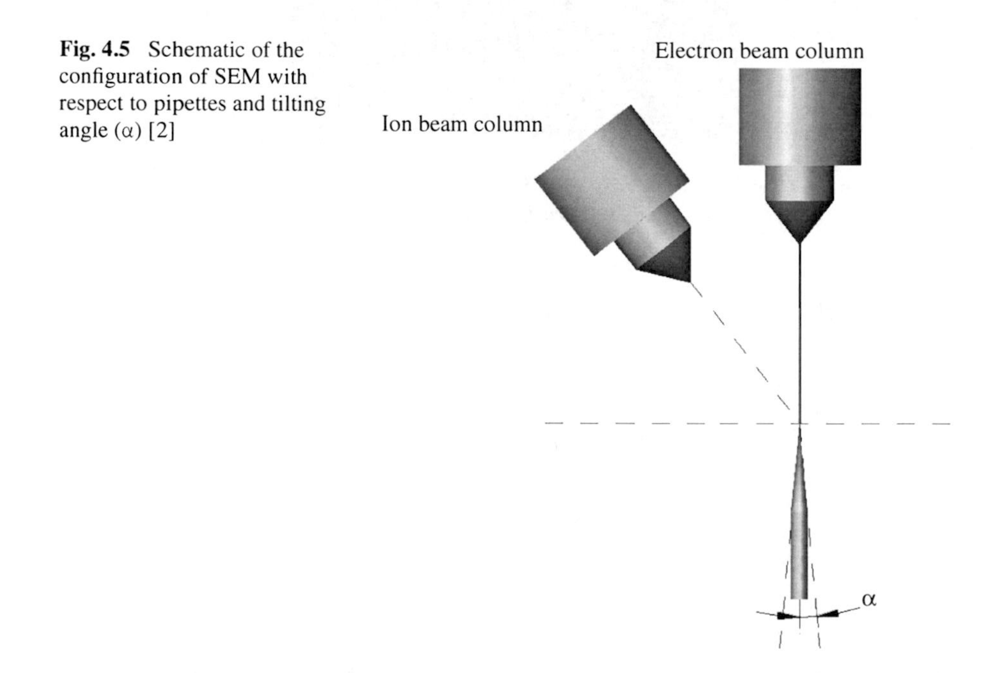

Fig. 4.5 Schematic of the configuration of SEM with respect to pipettes and tilting angle (α) [2]

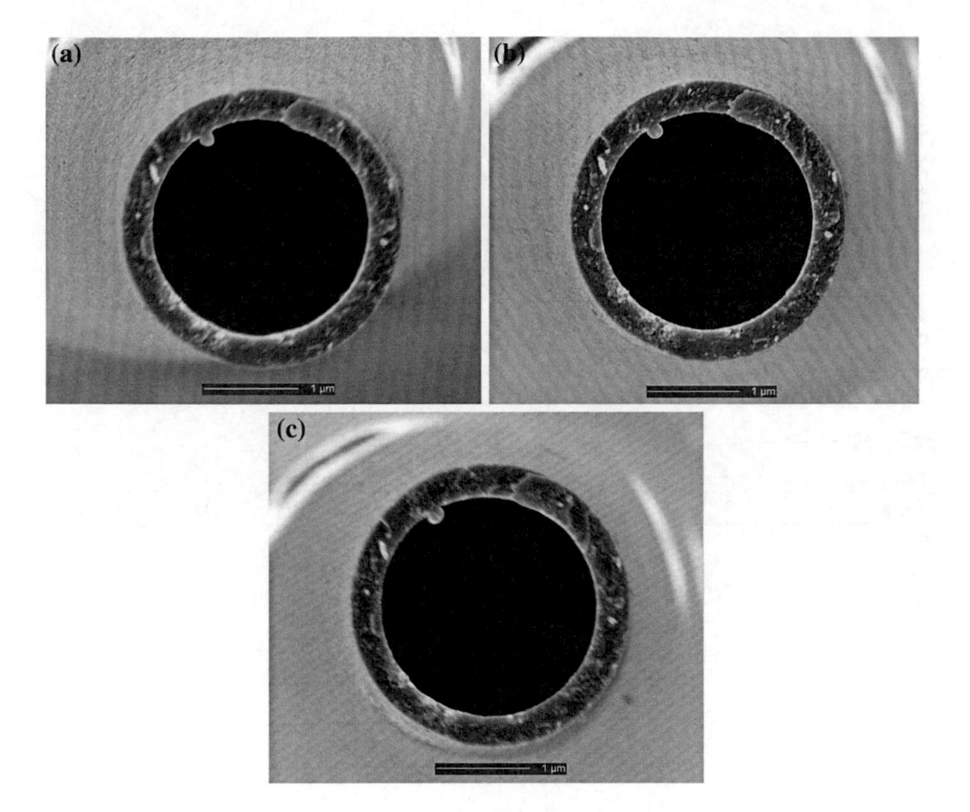

Fig. 4.6 Stereo images of the pipette tip for 3D reconstruction: **a** *left*, **b** *middle* and **c** *right*

of pipettes and electron beam gun with respect to each other. The SEM machine used for 3D reconstructions is a Strata DB 235 from FEI.

Three SEM images were taken from different angles by tilting the stage with respect to the electron beam direction. Figure 4.6a–c show the SEM images taken from the left, middle and right viewpoints of the pipette. The tilting angle between (a–b) and (b–c) of the images is 9 degrees.

The three SEM images were imported to the MeX software (a software package specialized in 3D reconstruction from SEM images) [16] and the digital elevation model (DEM) of the tip was obtained. Figure 4.7 shows the 3D reconstructed surface of the pipette tip.

Table 4.1 shows surface parameters of the 3D reconstructed pipettes with different sizes.

To have the highest lateral and vertical resolution in the 3D reconstructed surface, magnification, tilting angle and resolution should be as high as possible when capturing SEM images. Since the maximum pixel resolution of the machine is limited, different magnifications and tilting angles have been used to reconstruct each pipette's tip with the highest possible resolution. Such a reconstruction could

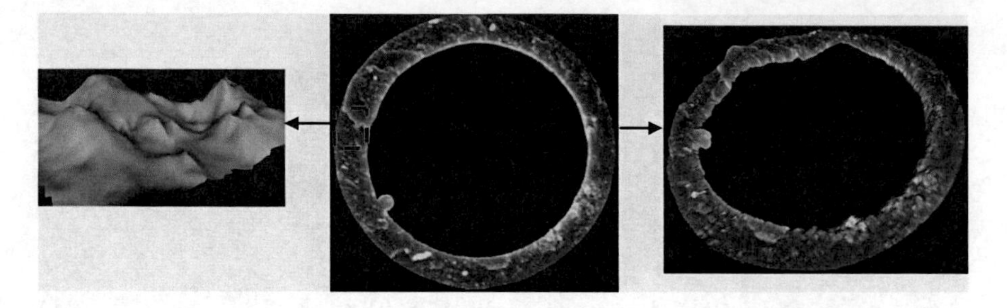

Fig. 4.7 A 3D reconstructed surface of the pipette tip shown at different viewing angles: *top view* (*middle*), exploded view of the area shown by *dash-line* (*left*), angled view (*right*)

Table 4.1 Surface parameters of pipette tip for 3 pipettes (D_t = tip size)

Name	Value $D_t = 2.2\ \mu m$	Value $D_t = 1.7\ \mu m$	Value $D_t = 1.3\ \mu m$	Description
S_a	27.3 nm	17.8 nm	8.3 nm	Average height of selected area
S_q	34.6 nm	13.0 nm	10.8 nm	Root-mean-square height of selected area
S_p	104.0 nm	81.5 nm	46.6 nm	Maximum peak of selected area
S_v	150.8 nm	125.5 nm	60.8 nm	Maximum valley depth of selected area
S_z	255.8 nm	207.1 nm	107.5 nm	Maximum height of selected area
S_{10z}	195.2 nm	142.2 nm	88.5 nm	Ten point height of selected area
S_{sk}	−0.225	−0.4857	−0.7099	Skewness of selected area
S_{ku}	3.2623	3.7725	4.5173	Kurtosis of selected area
S_{dq}	0.8774	0.9814	0.8289	Root mean square gradient
S_{dr}	34.986 %	45.575 %	32.61 %	Developed interfacial area ratio

Table 4.2 Reconstruction information for 3 pipettes

Pipette number	Tip size (μm)	Tilting angle (left to right)	Magnification	Lateral resolution (nm)	Vertical resolution (nm)
1	2.2	10	65000	18.4	18.4
2	1.7	10	95000	6.2	11.2
3	1.3	10	110000	5.3	7.7

be expected to have an inaccuracy of less than 5 % [14]. Table 4.2 gives the values of tip size, tilting angle, magnification, lateral resolution and vertical resolution of the several 3D reconstructed pipettes.

4.3 Polishing Pipette Tips by Focused Ion Beam Milling

As can be understood from Table 4.1, the surface roughness of the pipette tip is comparable with the thickness of the cell membrane which is 3 to 10 nm [17–20]. In order to study the effect of roughness on gigaseal formation two kinds of pipettes with distinct surface properties are required; rough pipettes (which are pipettes normally pulled by a pulling and heating process) and perfectly smooth pipettes. Focused ion beam (FIB) milling was used to produce pipettes with ideal tip surface conditions. The uneven surface of the pipette tip was corrected by cutting the top of the pipette across, using the Strata DB 235, dual-beam SEM/FIB system (FEI, Hillsboro, Oregon). The configuration of the pipette with respect to the ion beam is illustrated in Fig. 4.8.

Due to the conic shape of the pipette, cutting the tip changes the tip size, which is an important factor in patch clamping as it determines the pipette resistance [5, 21, 22]. It is also well known that a gigaseal is not likely to be achieved with large tip sizes [23–27] (see Chap. 6), so care was taken not to cut more than 1 μm from the top. Since the roughness of the tip of the pipette was in nanometres, cutting 1 μm from the top should be sufficient to remove all rough edges without increasing the tip size significantly. In the FIB milling process, the pipettes' tips were cut using Ga^+ ions with 50 pA current for 100 s, dwell time of 1μs and FIB acceleration voltage of 30 kV. The pipette before and after milling is shown in Fig. 4.9. The image of the milled pipette, shown in Fig. 4.9b, has a resolution of 4.5 nm. No feature could be identified on the milled surface for producing roughness parameters at this magnification. Therefore, the surface roughness of the milled pipette tip should be considerably less than 4.5 nm.

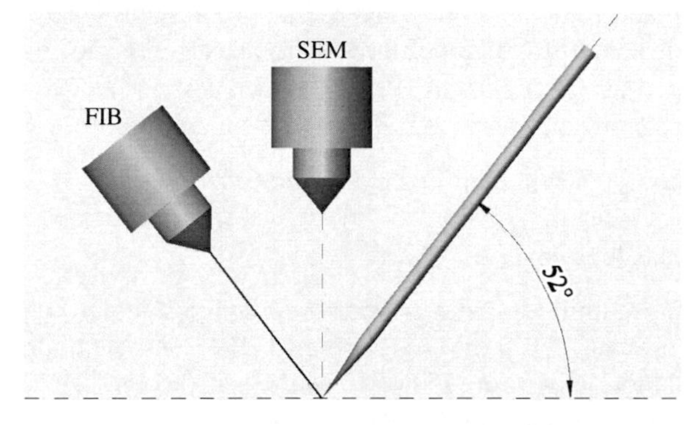

Fig. 4.8 The configuration of glass micropipette milling in the SEM/FIB chamber. The stage was tilted by 52 degrees so that the ion beam was perpendicular to the pipettes [2]

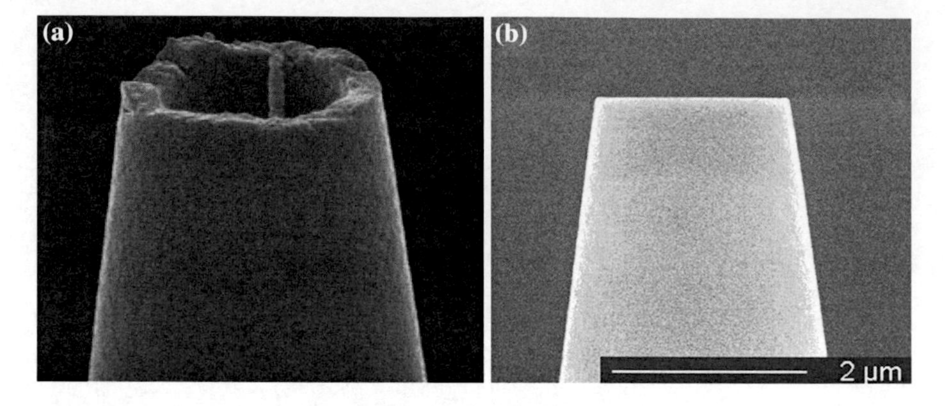

Fig. 4.9 FIB milling of a glass micropipette, **a** the micropipette before milling, **b** the pipette after the milling. No surface roughness could be identified after milling, so the surface roughness should be smaller than the resolution of the SEM image, which is 4.5 nm [2]

As was mentioned in Chap 3, an important step after pipette pulling is fire polishing. The process of FIB milling the pipette is called 'FIB polishing' in comparison with the 'fire polishing method'. FIB polishing has some advantages over fire polishing. Firstly, there is more control over the polishing process if using FIB, and the pipette tip can be polished at nanoscale without changing the tip shape; secondly, fire polishing has a blunting effect on the tip, which increases the contamination probability; and thirdly, FIB polishing does not change the sharpness of the pipette, which facilitates the pipette's travelling through tissues to the desired cells.

4.4 Patch Clamping Experiments

To investigate the effect of the roughness of pipette tips, patch clamp experiments were carried out with polished and conventional pipettes under the same conditions and the results were compared. Human Embryonic Kidney (HEK) cells were utilized to investigate the performance of the FIB-polished micropipettes in achieving gigaseals. The cells were cultured on cover slips in HEK cells medium, one to two days before the experiment, and incubation was done at 37 °C. The culture medium contained:

1. DMEM (Dulbecco's Modified Eagle's medium) 89 %
2. FBS (Fetal bovine serum) 10 %
3. PS (Penicillin Streptomycin) 1 %.

The backfilling solution (pipette solution) was composed of 40 mM KCl, 96 mM K-gluconate, 4 mM K2ATP, 2 mM GTP, 10 mM HEPE and with a pH value of 7.2, and the bath solution was composed of 110 mM NaCl, 5 mM KCl, 1 mM $MgCl_2$, 1 mM $CaCl_2$, 5 mM HEPEs, 5 mM HEPE–Na, and with a pH value of 7.2.

The experimental equipment setup consisted of an Axon multiclamp 700B microelectrode amplifier (Axon Instruments, Union City, CA), Flaming/Brown micropipette puller (Model P-97, Sutter Instruments) and glass micropipettes (BF150-86-10, Sutter Instruments). The puller machine was set to produce pipettes with a tip size of approximately 1.5 μm. When there was no contact between the recording pipette and the cell membrane, the total pipette resistance ranged from 6.0 to 6.5 MΩ. A 10 mV pulse was constantly applied to the recording electrode from the time that the pipette tip was just immersed in the bath solution till it touched the cell membrane. At this point the positive pressure were switched to suction to encourage seal formation (Fig. 4.10).

Ten recordings were obtained for each type of pipette. Seal resistances are shown in Fig. 4.11. With the FIB-polished pipettes, above 3 GΩ seals were

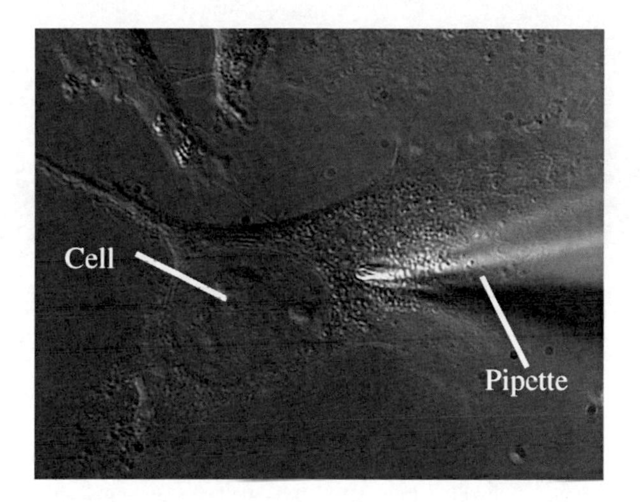

Fig. 4.10 Image of the *pipette* with respect to the *cell* at the moment of applying suction

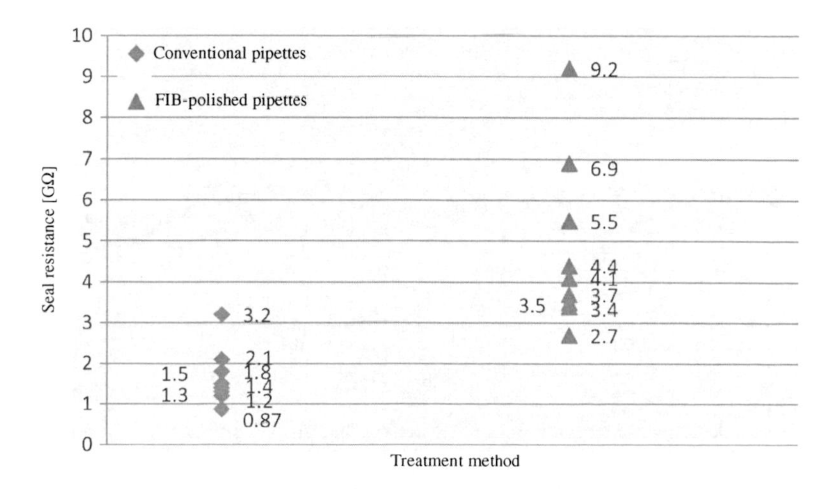

Fig. 4.11 Seal values for conventional and polished pipettes

achieved in most attempts and the highest seal resistance reached 9 GΩ. The mean value of seal resistances is 4.7 GΩ with a standard deviation of 1.8 GΩ. In comparison, the seal resistances achieved using the conventional pipettes were usually between 1.0 and 2.0 GΩ. The seal resistance could reach 3 GΩ in some excellent cases. The mean value of seal resistances is 1.6 GΩ with a standard deviation of 0.6 GΩ.

FIB-polished pipettes formed significantly better seals which made it possible to measure single ion channel currents with considerably lower noise. Single-channel currents recorded from conventional and polished pipettes are shown in Figs. 4.12 and 4.13.

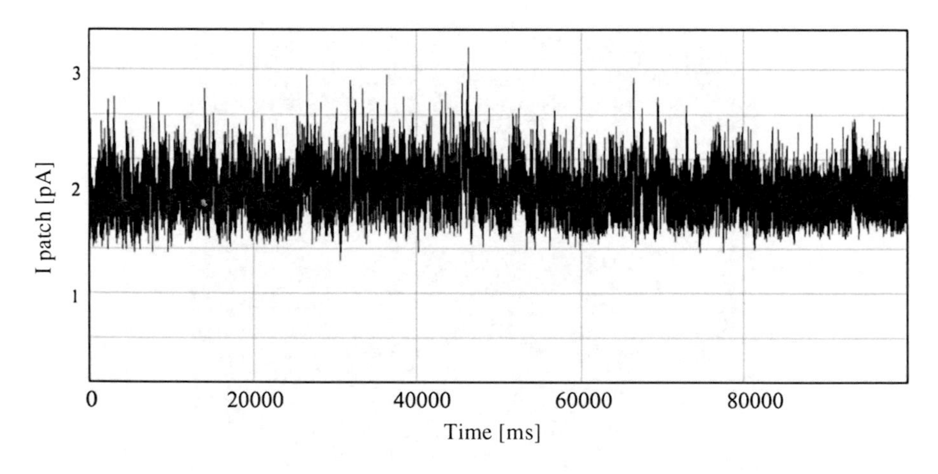

Fig. 4.12 Single-channel recording from HEK cells with conventional pipettes. The leakage current is 2.1 pA

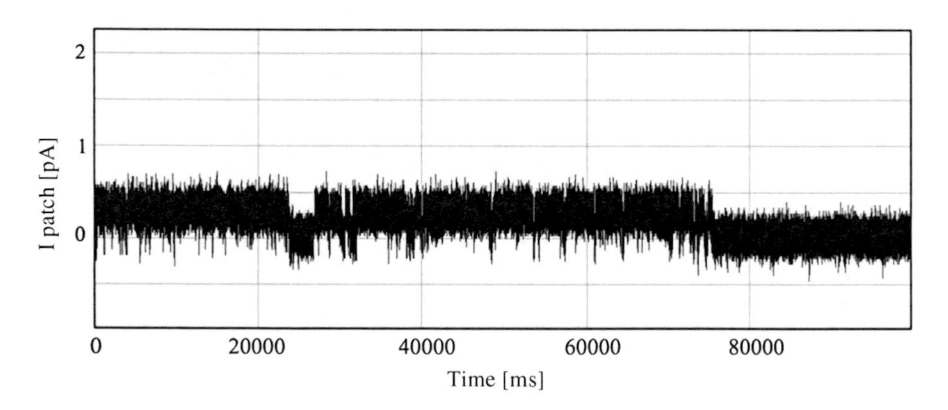

Fig. 4.13 Single-channel recording from HEK cells with polished pipettes. The improved patch clamping performance with polished pipettes is obtained from the better contact conditions of the smoother tip surface with the membrane. The leakage current is 0.3 pA, significantly lower than the current for conventional pipettes

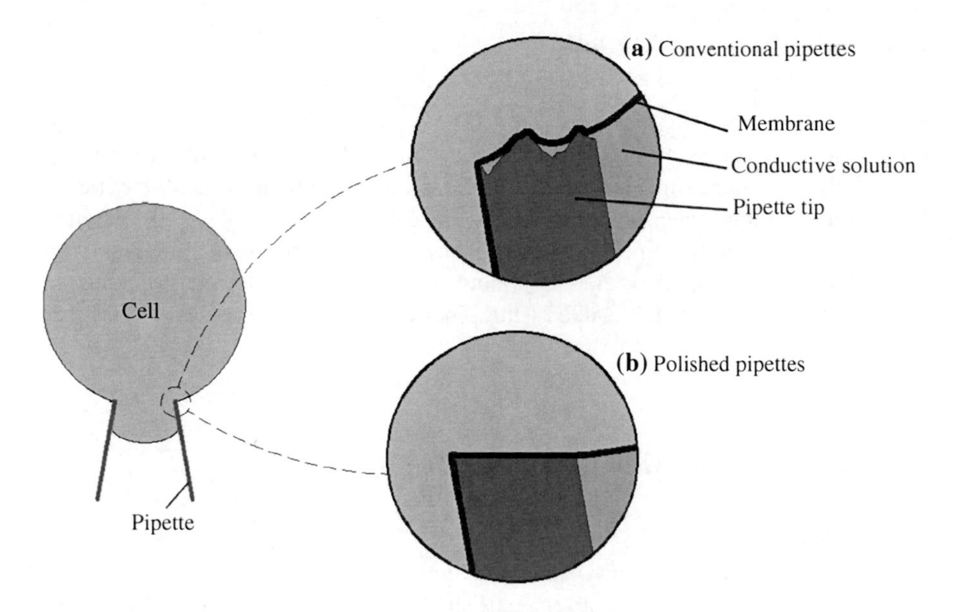

Fig. 4.14 Schematic of pipette-membrane interaction: **a** the original pipette tip with an uneven surface, **b** with a flat tip [2]

The higher seal resistance for polished pipettes could be explained by their better sealing potential. The perfectly smooth and flat surface of the polished pipette tip leaves no concave area to hold water, as opposed to conventional pipettes (Fig. 4.14). Contact area between pipette tip and cell membrane is higher for polished pipettes and since there are no peaks or spikes, the membrane can get closer to the tip. As a result it is more difficult for ions to escape from glass-membrane distance and higher seals are achievable. This also implies that the cell cannot fill the valleys of the rough surface of the conventional pipettes perfectly, which could possibly be the reason for reports of lower seal resistance with rough surfaces in the literature.

The results demonstrate two important things about the nature of gigaseal formation:

1. Although the seal is manly formed between the inner wall of the pipette and the cell membrane, the effect of the tip cannot be neglected. A flat, smooth tip can significantly decrease leakage.
2. The results show the importance of the closeness of the membrane to the glass over the length of opposition. The tip area is small compared to the inner wall area, yet the smooth flat tip can increase the seal value.

As discussed in Chap. 3, the seal forms between lipids and the inner wall of glass micropipettes. These findings show that the mechanism of gigaseal formation is not completely understood and there is no single model that can explain all observations.

4.5 Finite Element Modelling

Finite element modelling is carried out to study the effect of pipette tip roughness on gigaseal formation. How the cell deforms under the rough tip of a pipette has significant importance in gigaseal formation. If the membrane can fill cavities of the rough tip then higher seal resistances are expected due to the higher contact area between the rough pipette and the cell membrane. However if the membrane cannot fill the cavities then the space between two peaks acts like a conductive channel which connects the inside of the pipette to the outside. FE modelling was carried out using Abaqus/CAE software [28].

4.5.1 Patch Clamp Manipulators

Patch clamping involves the placement of a glass micropipette onto a cell to form a tight seal. The core function of a micromanipulator in patch clamping set-up is to place the micropipette tip onto the cell surface in a controlled way. The manipulator used in the experiments is MP-225 from Sutter Instruments (Fig. 4.15), which has a resolution of 62.5 nm for fine movements. As a result the pipette approaches the cell in a step of 62.5 nm. This implies that when a contact is made, the pipette tip is just on the cell surface or presses the cell within tens of nanometres. If the pipette continues its movement, it may penetrate or rupture the cell membrane. Figure 4.16 shows the right moment of applying suction.

In practice the relative position of tip to membrane is estimated by monitoring the change of electrical resistance between the two electrodes. As the pipette comes closer to the membrane the resistance increases; usually an increase of about 1 MΩ indicates that the tip has touched the membrane [21]. It is not recommended to reach a resistance 1.5 times higher than the pipette resistance as it stresses the membrane and contaminates the tip, which is fatal for gigaseal formation. In finite element modelling, once in contact, the pipette is lowered by 500 nm and the suction is then applied.

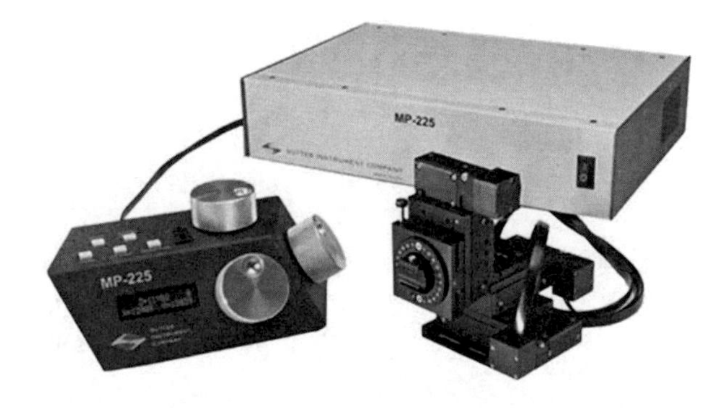

Fig. 4.15 MP-225 micromanipulator from Sutter Instruments [10]

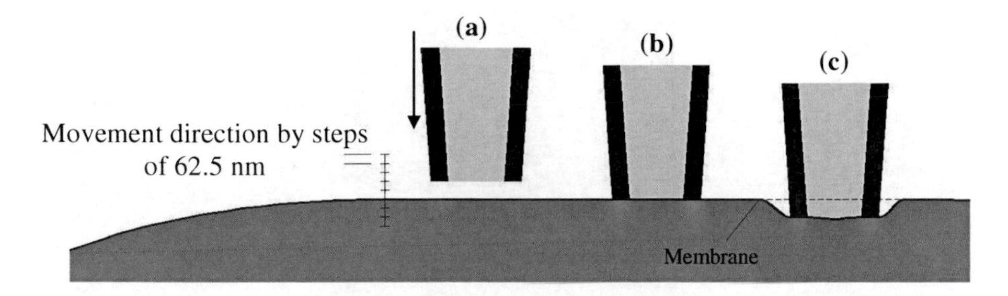

Fig. 4.16 The pipette approaches the cell by steps of as long as manipulator resolution. **a** The pipette is a long way from the membrane, and application of suction at this point may increase the resistance, but usually the resistance drops after suction is released and the tip should be considered contaminated, therefore needing to be changed. **b** The pipette is on the membrane; this is the right position to apply suction for most of the cells, and **c** the pipette pushes in the membrane by a few tens of nanometres; this is the most usual case in patch clamping. More pressing may result in membrane rupture; however, for some cell types the pipette needs to be pressed against the cell membrane more strongly [29]

4.5.2 Pipette Tip Profile

Figure 4.17 shows the digital elevation model image of the pipette used in FE modelling with four different profiles (numbered 1–4) of the tip surface along its thickness. The specifications of the profiles are given in Table 4.3. In order for the modelled pipette to have the real tip profile of the glass micropipette, coordinates of each profile are extracted and a B-spline is drawn through these points. The splines are then transferred to ABAQUS/CAE for finite element simulation. The inner and outer diameters of the pipette tip are 0.7 μm and 1 μm, respectively. As can be seen in Fig. 4.17, there is a large variation in the surface morphology of the pipette tip across its thickness. Therefore different tip profiles were used in the modelling.

4.5.3 Finite Element Modelling of Patch Clamping

There are three elements in this modelling: the glass micropipette, substrate and the cell. The pipette and substrate are modelled as 2D elastic solid bodies.

Two different mechanical models can be applied to cells, and either one allows the simulation of cellular deformations in response to micropipette aspiration in which the cell is sucked into the micropipette by applying negative pressure. The first one describes cells as having a solid membrane and liquid core and has been used to model the aspiration of cells with little or no cytoskeleton (such as red blood cells or unattached leukocytes) into micropipettes. The second one describes cells spread on a substrate (which is the case in conventional patch clamping) with a well developed cytoskeleton as being elastic solids [18, 30–32]. Cells can be modelled as continuum media if the smallest operative length scale of interest

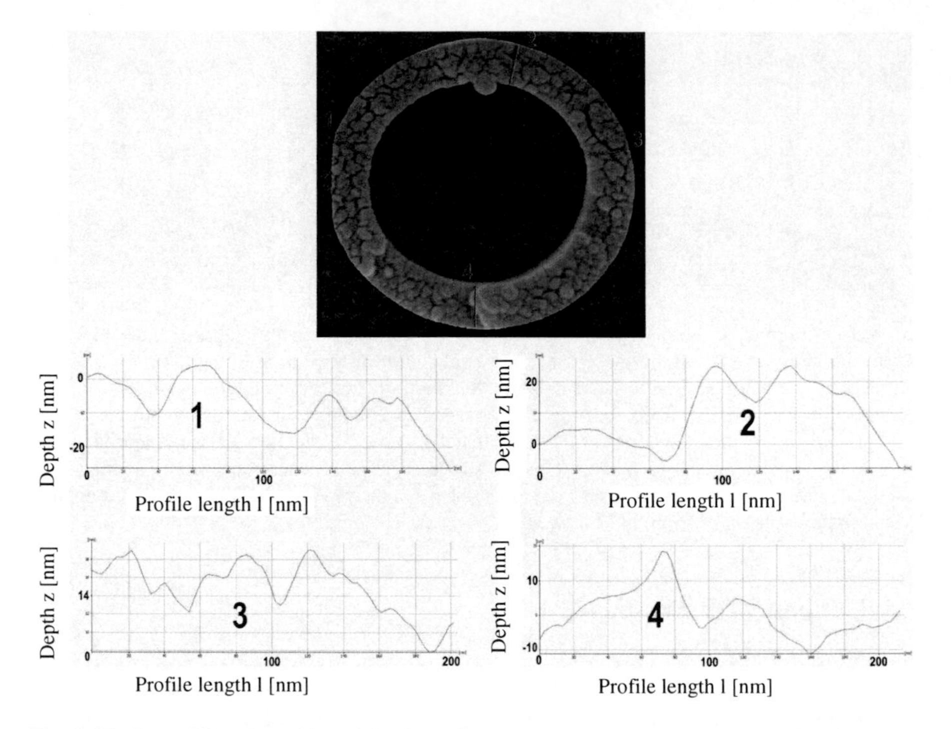

Fig. 4.17 Four different profiles of the tip surface across the thickness are shown. The large variation of surface morphology compromises the formation of a high-resistance seal

is significantly larger than the distance over which cellular structure or properties may vary [33–35]. Here, the cell is modelled as a continuous, homogeneous, incompressible, isotropic and hyperelastic solid attached to the substrate. The modelled cell consists of two parts: the cytoplasm and the membrane. Table 4.4 gives the material properties of cell membrane and cytoplasm. The material properties of cytoplasm and membrane used in this modelling are: $E_c = 1000$ Pa, $\nu_c = 0.5$, $\rho_c = 1000$ kg/m^3, $E_m = 100$ Mpa, $\nu_m = 0.3$, $\rho_m = 1150$ kg/m^3.

Since the physical combination of the pipette and cell is completely symmetric, a 2D axisymmetric model is employed instead of using a huge 3D model with a large number of nodes and elements. The cytoplasm is modelled as a semicircle meshed with 4-node bilinear axisymmetric quadrilateral elements. The plasma membrane is modelled as a thin elastic shell. Dimensions and boundary conditions are shown in Fig. 4.18. In the finite element simulation the pipette is lowered for 300 nm and then the suction is applied to the inside of the pipette. The membrane is tethered to the cytoplasm using a tie constraint. Contact between pipette and cell is considered frictionless.

The result of the finite element modelling is shown in Fig. 4.19. Although the cytoplasm is soft, the stiffer membrane does not allow the cell to cover all of the cavities of the tip. The highest peaks of the profile push down the membrane. Therefore, the distance between two peaks will not be filled with the patched

Table 4.3 Profile parameters of four different profiles

Profile	No.1	No.2	No.3	No.4	Description
P_a	4.0 nm	6.2 nm	2.1 nm	4.8 nm	Average height of profile
P_q	4.8 nm	7.8 nm	2.3 nm	6.2 nm	Root-mean-square height of profile
P_t	22.1 nm	34.0 nm	9.2 nm	29.2 nm	Maximum peak to valley height of primary profile
P_z	13.0 nm	12.4 nm	5.5 nm	12.6 nm	Mean peak to valley height of primary profile
P_{max}	18.4 nm	20.8 nm	6.7 nm	16.5 nm	Maximum peak to valley height of primary profile within a sampling length
P_p	8.7 nm	14.7 nm	4.8 nm	16.3 nm	Maximum peak height of primary profile
P_v	13.4 nm	19.3 nm	4.4 nm	12.9 nm	Maximum valley height of primary profile
P_c	14.8 nm	26.1 nm	5.4 nm	19.7 nm	Mean height of profile irregularities of primary profile
P_{sm}	92.7 nm	150.54 nm	49.7 nm	76.6 nm	Mean spacing of profile irregularities of primary profile
P_{sk}	−0.741	−0.2124	0.08	0.1866	Skewness of primary profile
P_{ku}	2.4831	2.5544	2.272	3.0219	Kurtosis of primary profile
P_{dq}	0.4716	0.6446	0.279	0.5801	Root-mean-square slope of primary profile

Table 4.4 Material properties of cytoplasm and membrane

	Young modulus (E)	Poisson's ratio (ν)	Density (ρ) kg/m^3
Cytoplasm	75–2500 Pa [20, 31, 32, 36]	0.5 [19, 31, 37]	1000 [19, 20, 38]
Membrane	100–140 MPa [19, 20]	0.3–0.5 [18, 19, 20]	1150 [20]

membrane. This is important because if the membrane fills up all of the room between peaks and valleys, then the contact area between the membrane and the glass will increase. This in turn can result in a better seal. However the result of the FE modelling shows that the membrane cannot go into the valleys. Therefore the spaces, which are not filled by the cell, act like channels connecting the inside and outside of the pipette together. These channels are filled with the conductive media, making it easier for ions to escape, therefore increasing current leakage and compromising the seal. As can be understood from Table 4.3, the maximum peak to valley height of these channels is about 10–34 nm. The result of the FE

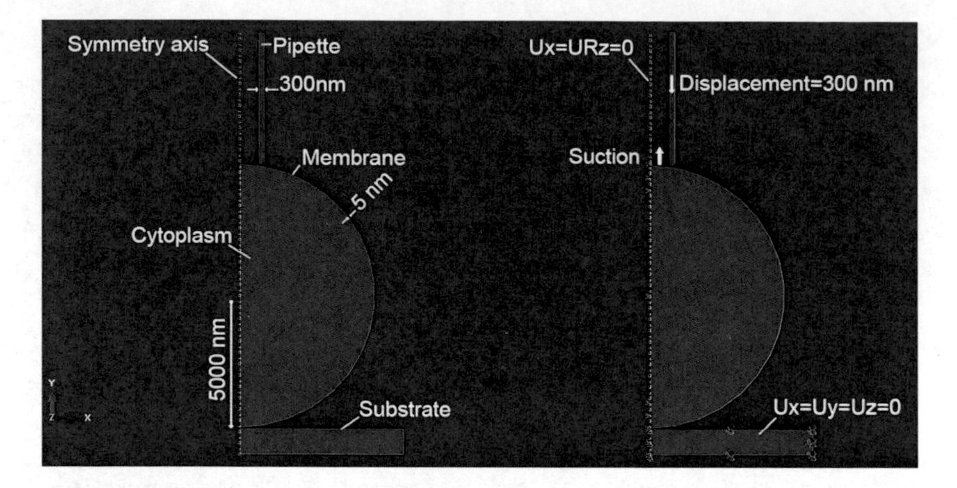

Fig. 4.18 Dimensions and boundary conditions used in the FE modelling. Nodes on the symmetry axis are prevented from movement in *X* direction and rotation around *Z* axis. The substrate is not allowed to have any movement in any direction

modelling is in agreement with experiments carried out with rough and polished pipettes. Polished pipettes make a better seal with cells because their flat tip can make a better contact with the cell surface.

4.6 Summary

In this chapter the effect of pipette tip roughness on gigaseal formation in patch clamping has been studied. Micropipettes are fabricated in a heating and pulling process. High-magnification SEM images of pipettes' tips have shown that they are rough and jagged. The SEM stereoscopic technique has been used for 3D reconstruction of the pipette tip and roughness parameters were extracted from digital elevation models (DEM) of the tips. FIB milling is used to cut across the tips, leaving a very smooth surface at the top of the pipettes. Patch clamping experiments were carried out using FIB-polished and conventional pipettes. Seal values are considerably higher in the case of polished pipettes. Above 3 GΩ seals were achieved readily and the highest seal resistance reached was 9 GΩ for polished pipettes. The leakage current in single-channel recording was found to be 0.3 pA, significantly smaller than the 2–3 pA usually achieved using conventionally treated pipettes. The smaller current is the consequence of higher seal resistance. To further investigate the effect of roughness on gigaseal formation FE modelling of patch clamping was carried out. The results of the FE modelling show that the cell cannot fill up all of the valleys of the tip and therefore in three-dimensions the inside of the pipette is connected to the outside by nanometre-high channels, facilitating current leakage. This is consistent with the result obtained from patch clamping experiments, where

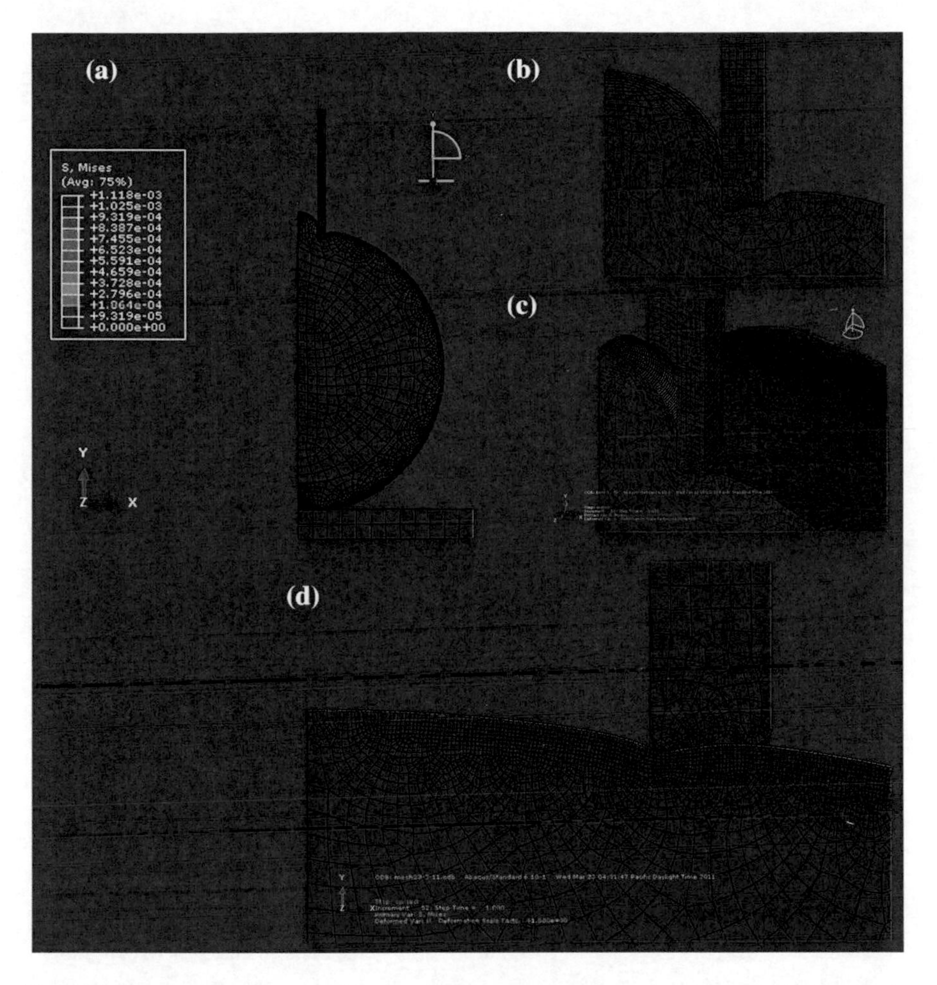

Fig. 4.19 Result of the FE modelling of patch clamping process, **a** Aspiration of cell into pipette, **b** close view of the region underneath the pipette. Stiff cell membrane prevents soft cytoplasm from filling the cavities of the tip, **c** 3D representation of result, **d** cell deformation at the end of indentation process and before applying suction. The material properties of cytoplasm and membrane used in this modelling are: $E_c = 1000$ Pa, $\nu_c = 0.5$, $\rho_c = 1000$ kg/m^3, $E_m = 100$ MPa, $\nu_m = 0.3$, $\rho_m = 1150$ kg/m^3

smoother tips resulted in higher seal values. Thus, FIB-polished glass micropipettes have improved the gigaseal formation in patch clamping.

References

1. Malboubi M, Gu Y, Jiang K (2010) Study of the tip surface morphology of glass micropipettes and its effects on giga-seal formation. Electronic engineering and computing technology. Springer, pp 609–619
2. Malboubi M et al (2009) Effects of the surface morphology of pipette tip on giga-seal formation. Eng Lett 17:281–285

3. Malboubi M et al (2009) The effect of pipette tip roughness on giga-seal formation. World Congr Eng 2:1849–1852
4. Fertig N, Blick RH, Behrends JC (2002) Whole cell patch clamp recording performed on a planar glass chip. Biophys J 82:3056–3062
5. Li S, Lin L (2007) A single cell electrophysiological analysis device with embedded electrode. Sens Actuators A 134:20–26
6. Matthews B, Judy JW (2006) Design and fabrication of a micromachined planar patch-clamp substrate with integrated microfluidics for single-cell measurements. J Microelectromech Syst 15:241–252
7. Stett A et al. (2003) Cytocentering: A novel technique enabling automated cell-by-cell patch clamping with the Cytopatchtm chip. Recept Channels 9:59–66
8. Ong WL, Yobas L, Ong WY (2006) A missing factor in chip-based patch clamp assay: gigaseal. J Phys 34:187–191
9. Operational Manual P-97 Flaming/Brown Micropipette Puller Sutter instrument company. [Online] www.sutter.com
10. [Online] Sutter Instruments. www.sutter.com
11. Piazzesi G (1973) Photogrammetry with the scanning electron microscope. J Phys E: Sci Instrum 6:392–396
12. Bariani P et al (2005) Investigation on the traceability of three dimensional scanning electron microscope measurements based on the stereo-pair technique. Precis Eng 29:219–228
13. Samak D, Fischer A, Rittel D (2007) 3D reconstruction and visualization of microstructure surfaces from 2D images. J Manuf Technol 56:149–152
14. Marinello F et al (2008) Critical factors in SEM 3D stereomicroscopy. Meas Sci Technol 19:1–12
15. MeX catalouge. Graz, Austria: Alicona Imaging GmbH
16. MeXTM software. Graz, Austria: Alicona Imaging GmbH
17. Guyton AC, Hall JE (2000) Text book of medical physiology. Saunders company, 0-7216-8677-X
18. Charras GT et al (2004) Estimating the sensitivity of mechanosensitive ion channels to membrane strain and tension. Biophys J 87:2870–2884
19. Tang Y et al. (2009) Numerical simulation of nanoindentation and patch clamp experiments on mechanosensitive channels of large conductance in Escherichia coli. Exp Mech 49:35–46
20. Bae C, Butler PJ (2008) Finite element analysis of microelectrotension of cell membranes. Biomech Model Mechanobiol 7:379–386
21. Molleman A (2003) Patch clamping: an introductory guide to patch clamp electrophysiology. Wiley, Chichester 0-471-48685-X
22. Kornreich BG (2007) The patch clamp technique: principles and technical considerations. J Vet Cardiol 9:25–37
23. Klemic K, Klemic J, Sigworth F (2005) An air-molding technique for fabricating PDMS planar patch-clamp electrodes. Eur J Physiol 449:564–572
24. Kusterer J et al (2005) A diamond-on-silicon patch-clamp-system. Diamond Relat Mater 14:2139–2142
25. Sakmann B, Neher E (2009) Single-channel recording. 2nd edn. Springer, New York, 978-1-4419-1230-5
26. Petrov AG (2001) Flexoelectricity of model and living membranes. Biochimica et Biophysica Acta 1561:1–25
27. Petrov AG (2006) Electricity and mechanics of biomembrane systems: Flexoelectricity in living membranes. Analytica Chimica Acta 568:70–83
28. ABAQUS (2004) ABAQUS 6.4 user's manual. ABAQUS Inc, Providence RI
29. Malboubi M, Gu Y, Jiang K (2009) Experimental and simulation study of the effect of pipette roughness on giga-seal formation in patch clamping. Microelectronic Eng 87:778781
30. Baaijens VPT et al. (2005) Large deformation finite element analysis of micropipette aspiration to determine the mechanical properties of the chondrocyte. Annal Biomed Eng 33:494–501

31. Caille N et al. (2002) Contribution of the nucleus to the mechanical properties of endothelial cells. J Biomech 35:177–187
32. Zhou EH, Lim CT (2005) Finite element simulation of the micropipette aspiration of a living cell undergoing large viscoelastic deformation. Mech Adv Mater Struct 12:501–512
33. Mofrad RK, Kamm RD (2006) Cytoskeletal mechanics. Cambridge university press, Cambridge, 978-0-521-84637-0
34. Vaziri A, Gopinath A, Deshpande V (2006) Continuum-based computational models for cell and nuclear mechanics. J Mech Mater Struct 2:1169–1191
35. Boal D (2001) Mechanics of the cell. Cambridge University Press, Cambridge, 978-0-521-79681-1
36. Yokokawa M, Takeyasu K, Yoshimura SH (2008) Mechanical properties of plasma membrane and nuclear envelope measured by scanning probe microscope. J Microsc 232:82–90
37. Theret DP et al (1988) The application of a homogeneous half-space model in the analysis of endothelial cell micropipette measurements. J Biomech Eng 110:190–199
38. Hartmann C, Delgado A (2004) Stress and strain in a yeast cell under high hydrostatic pressure. In: Proceedings in applied mathematics and mechanics, vol 4, pp 316–317

Chapter 5
Effect of Hydrophilicity on Gigaseal Formation

A membrane has both hydrophilic and hydrophobic components. The exact contribution of these components in seal formation is not clear. Hydrophilicity of the pipette (or patch site in planar patch clamping) is believed to be a prerequisite for gigaseal formation [1]. For many materials, treating the substrate to be more hydrophilic resulted in considerably higher seal values [2–5]. However the twofold structure of the membrane has made it possible for the membrane to form a seal with both hydrophobic and hydrophilic materials [6]. In this chapter the effect of hydrophilicity on gigaseal formation is discussed.

5.1 Piranha Solution Treatment

Piranha solution has been in use in the semiconductor industry for decades. The piranha solution is a mixture of concentrated sulphuric acid (H_2SO_4) and hydrogen peroxide (H_2O_2), used to remove organic contaminants from the surface. The solution is a strong oxidizer and will also make the surface extremely hydrophilic [7]. Both cleanliness and hydrophilicity are important in seal formation. Many different mixture ratios are commonly used, but a typical piranha solution consists of the following: 3:1 vol/vol 96 % H_2SO_4:30 % H_2O_2. A mixture of the two results in the formation of the strong oxidant H_2SO_5 [8, 9]:

$$H_2SO_4 + H_2O_2 \rightarrow H_2SO_5 + H_2O$$

As a result piranha solution is a strong oxidizer and will hydroxylate the surface by increasing silanol groups and Si–O-species on the glass, making the surface more hydrophilic [10].

Glass cover slips were used for characterization of the treatment procedure before treating the micropipettes. Cover slips were dipped in piranha solution for 30 min and the temperature was kept at 85 °C to maintain the effectiveness of the solution. Figure 5.1 shows the contact angle between the treated glass and water before and after treatment.

M. Malboubi and K. Jiang, *Gigaseal Formation in Patch Clamping*,
SpringerBriefs in Applied Sciences and Technology,
DOI: 10.1007/978-3-642-39128-6_5,

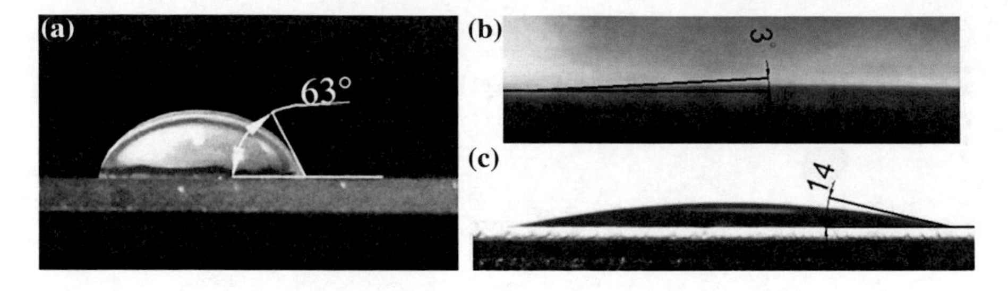

Fig. 5.1 Piranha solution treatment of glass cover slips for 30 min. **a** Contact angle between water and glass before treatment, **b** after treatment and **c** 2 h after treatment

Treating micropipettes is different from treating cover slips. In order to apply piranha treatment to glass micropipettes the following issues should be considered carefully:

- Treatment time
- Effect of piranha treatment on pipette capacitance and
- Effect of piranha treatment on pipette surface roughness.

5.1.1 Treatment Time

Figure 5.1 shows that piranha solution treatment is an effective way to increase the hydrophilicity of glass. However, dipping pipettes in piranha solution for 30 min causes a considerable amount of piranha solution to be sucked into the pipette by capillary action. Piranha solution is harmful to cells and it must be removed from pipettes before conducting patch clamp experiments. Due to the small tip size of the pipettes (1–2 μm), it takes a long time to remove the piranha solution by applying pressure to the back of the pipettes. Shorter treatment times were used to overcome this problem by decreasing the amount of solution which goes into the pipettes. To observe the effectiveness of treatment, the contact angle between a water droplet and the glass cover slips was measured for different treatment times and is presented in Table 5.1. As patch clamp experiments were usually carried

Table 5.1 Contact angle for different treatment times

Treatment time (sec)	Contact angle measured after treatment	Contact angle measured after 2 h.
10	28	48
20	11	37
30	10	19

A treatment time of 30 s was chosen because it is more effective and the surface can maintain its properties for a longer time.

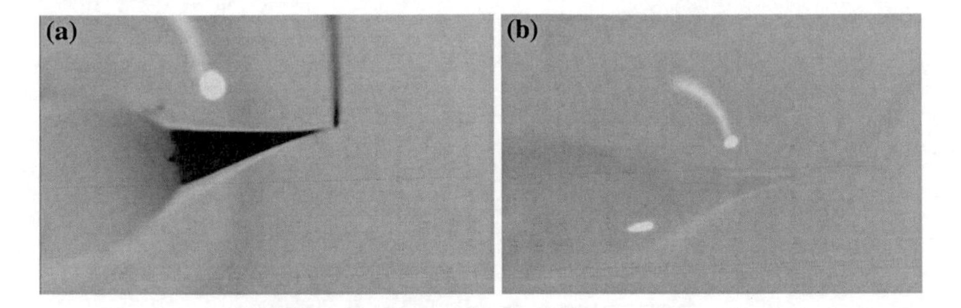

Fig. 5.2 Pipettes were dipped in ink for 30 s. **a** Positive pressure was applied to the pipette for 2 min and 40 s to remove the ink. The experiments were carried out under water to eliminate surface tension at the tip. **b** Ink has been completely removed from the pipette

out with a delay from treatment, contact angles are measured after treatment and after 2 h from treatment to find out if the surface is able to keep its properties for a period of time.

Based on the values of contact angles in Table 5.1, a treatment time of 30 s was chosen.

To estimate the influence of capillary action for this treatment time pipettes were dipped in ink for 30 s. This gives a good approximation of the length of time required for applying pressure to the pipette to remove the piranha solution after treatment. Blue ink was used to give a higher contrast as piranha solution is transparent and it is difficult to observe it leaving the pipette. It took 2 min and 40 s to completely remove the ink from the pipette which was sucked in by capillary action in 30 s (Fig. 5.2). Before conducting patch clamp experiments positive pressure is applied to pipettes for 3 min to remove the piranha solution.

5.1.2 Effect of Piranha Solution Treatment on Pipette Surface Roughness

Piranha solution etches glass. Seu et al. have measured the surface roughness of glass slips treated with piranha solution for different lengths of time [10]. Figure 5.3 shows the surface roughness values of glass after various treatment times. Noticeable roughening can be observed as the etch time is increased.

5.1.3 Effect of Piranha Treatment on Pipette Capacitance

Pipette capacitance is an important factor in patch clamp recordings and should be minimized. Treating pipettes with piranha solution increases the hydrophilicity of

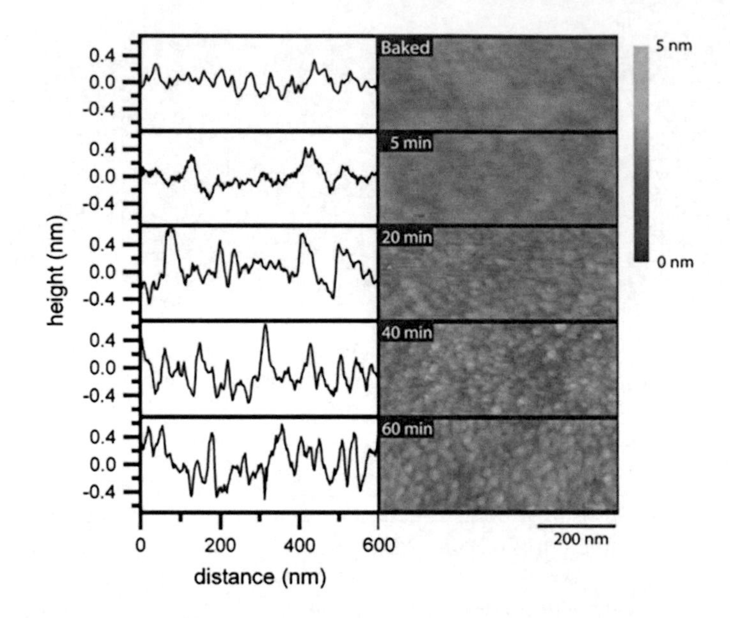

Fig. 5.3 Effect of piranha solution etching time on surface roughness of glass slides. AFM images and line scans from five different samples are shown. Surface roughness increases with etching time [10]. Decreasing treatment time from several minutes to several seconds will greatly reduce etching. Treating pipettes for 30 s with piranha solution should have a negligible effect on the surface roughness of glass micropipettes

the glass surface and facilitates the creeping up of conductive watery solution from the pipette wall. This increases the pipette capacitance. Therefore only the very end of pipettes should be treated and the rest of the pipettes should be preserved from treatment. This has been done by dipping only the end of micropipettes in piranha solution using a simple pipette holder (Fig. 5.4).

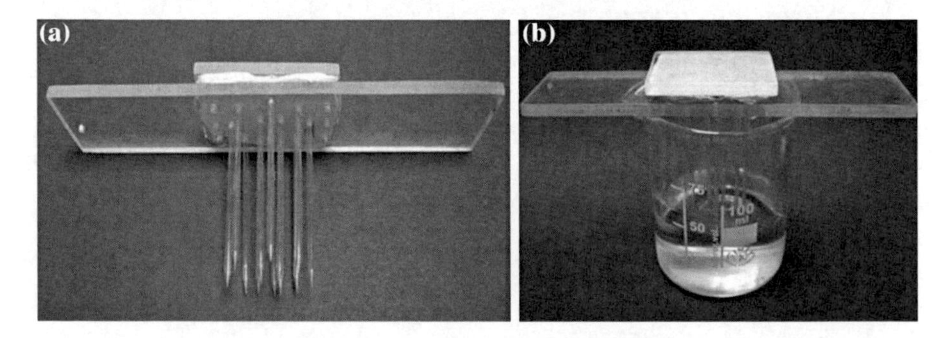

Fig. 5.4 Only the very ends of pipettes were dipped into piranha solution using a micropipette holder to prevent the rest of the micropipettes from being treated. **a** The pipette holder with pipettes assembled in the holder; **b** Pipettes were inserted into the piranha solution

5.1.4 Piranha Solution Treatment of Glass Micropipettes

The purpose of treating glass micropipettes with piranha solution is to increase the hydrophilicity of the inner wall of the micropipettes which is in contact with the cell membrane in seal formation. Pipettes were dipped in piranha solution for 30 s using the pipette holder. Then pipettes were backfilled with pipette solution and pressure is applied to them after immersing the tip into bath solution for 3 min. This time is enough to remove harmful solution from the tip. The pipettes are now ready for performing patch clamp experiments.

5.2 Oxygen Plasma Treatment

The second method used to alter the surface properties of glass micropipettes is oxygen plasma treatment. Plasma is a partially ionized gas with an equal number of positive and negative charges. The ions in the plasma are accelerated through the plasma sheet and bombard the surrounding surfaces. Plasma treatment increases hydrophilicity and cleans the surface. The plasma affects a surface physically (bombardment with energetic ions) and chemically (interactions of chemically active species in the plasma). Depending on plasma parameters (power, applied voltage, pressure, plasma density etc.) and gas chemistry, plasma discharges can be employed for etching, deposition or surface cleaning [11]. Oxygen plasma treatment leads to the formation of surface (–OH) groups (silanol) (Fig. 5.5). The increased concentration of OH groups at the surface provides a higher number of siloxane bonds. Glass cover slips were used for characterization of the treatment procedure. They were exposed to oxygen plasma for 1 min. Figure 5.6 shows the contact angle between the glass and a droplet of water before and after oxygen plasma treatment.

As with piranha solution treatment, pipette capacitance and surface roughness are important factors which need careful consideration in plasma treatment.

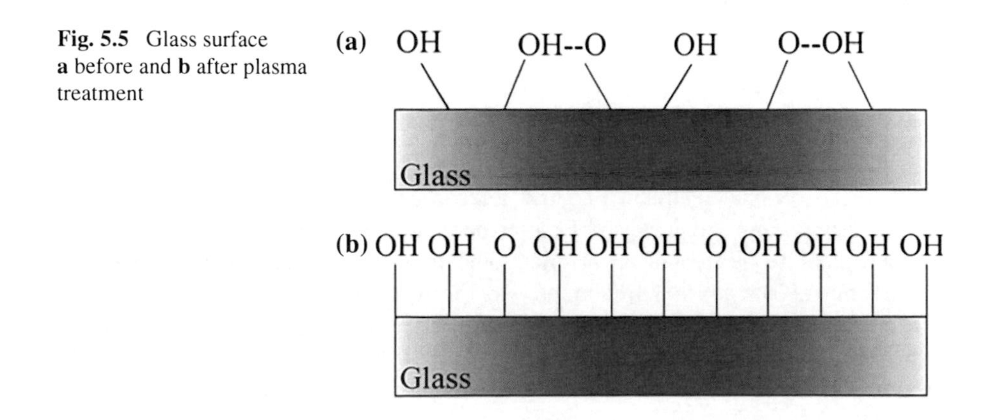

Fig. 5.5 Glass surface **a** before and **b** after plasma treatment

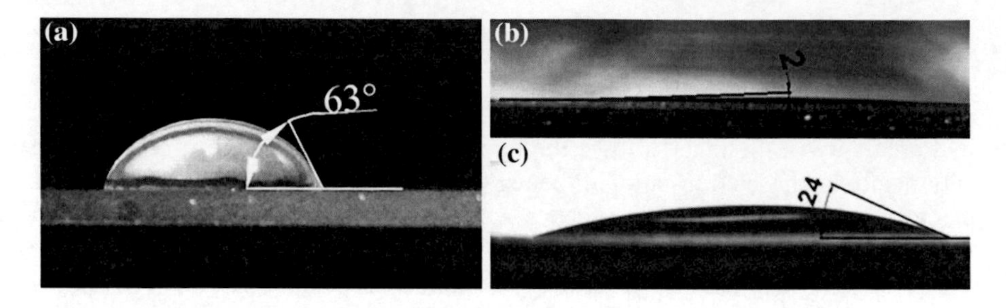

Fig. 5.6 Oxygen plasma treatment of glass coverslips for 1 min. **a** Contact angle between water and glass before treatment, **b** after treatment and **c** 2 h after treatment

5.2.1 Effect of Oxygen Plasma Treatment on Pipette Capacitance

As stated earlier, it is only the tips of micropipettes that need to be treated and the rest of the pipettes should be preserved from treatment. A holder was designed for this purpose. Before plasma treatment pipettes were assembled into the columns of the holder and the holder was placed into the plasma chamber.

5.2.2 Effect of Oxygen Plasma Treatment on Surface Roughness

Choi et al. showed that oxygen plasma treatment increases the surface roughness of glass [12]. The amount of roughness is proportional to the magnitude of power and exposure time. The oxygen radicals preferentially remove weak Si–Si bonds and break Si–O bonds at the surface, which results in higher surface roughness [12]. In treating glass micropipettes oxygen plasma parameters were well chosen to minimize the increase in surface roughness.

5.2.3 Oxygen Plasma Treatment of Glass Micropipettes

For oxygen plasma treatment of glass micropipettes, these were inserted into a low-pressure radio frequency (RF) plasma chamber. Ten pipettes were placed in the holder to be treated all at once. The holder was inserted horizontally to the chamber. The plasma treatment was performed at a working pressure of 40 mTorr, an oxygen flow of 50 sccm and 40°. The oxygen plasma power and exposure time were respectively, 800 W inductively coupled power, 20 W platen power, and 1 min. According to Choi et al. for these values the size of roughness is below 2 Å. This amount of change in surface roughness is negligible in comparison with the surface roughness of glass micropipettes (see Chap. 5). Therefore oxygen plasma treatment has a minimum effect on the surface properties of pipettes.

5.3 Patch Clamping Experiments

Patch clamping experiments were carried out on HEK (Human Embryonic Kidney) cells with piranha solution-treated, oxygen plasma-treated and conventional pipettes and the results are compared. Cell culture, materials and the set-up for patch clamping experiments were the same as those discussed in Chap. 4. Ten measurements were made for each type of pipette and seal values were recorded. The mean value of seal resistances for conventional pipettes is 1.6 GΩ with the standard deviation of 0.6 GΩ. The mean of seal resistances for piranha solution-treated pipettes is 3.0 GΩ with a standard deviation of 0.9 GΩ. In comparison, the mean value of seal resistance for oxygen plasma treated-pipettes is 0.93 GΩ with a standard deviation of 0.3 GΩ. Figures 5.7 and 5.8 show the performance of one of the piranha solution-treated and one of the oxygen plasma-treated pipettes respectively. Figure 5.9 shows the seal values for oxygen plasma-treated, piranha solution-treated and conventional pipettes.

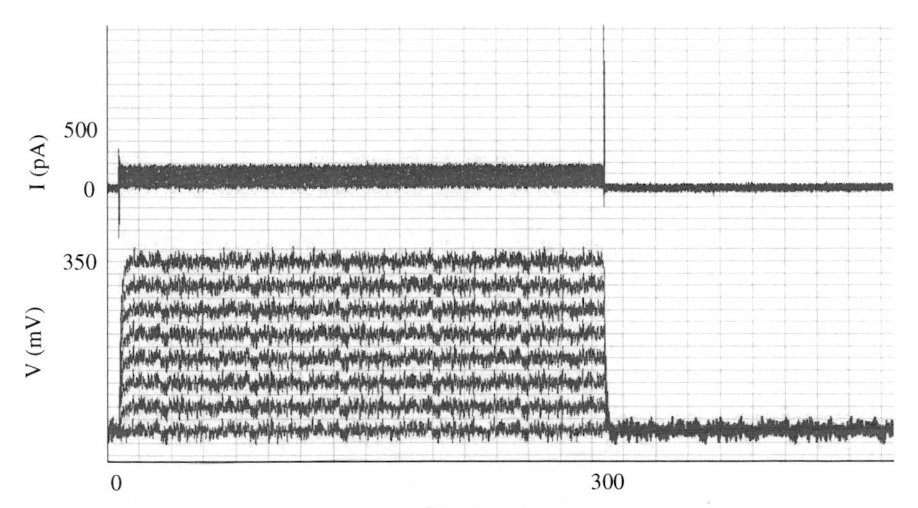

Fig. 5.7 Voltage clamp recordings showing changes in current performed by an oxygen plasma-treated pipette. The voltage step length is 30 ms, the increment is 50 mV per step. The application of a 350 mV pulse resulted in a recorded current of approximately 210 pA and a calculated seal resistance of 1.6 GΩ

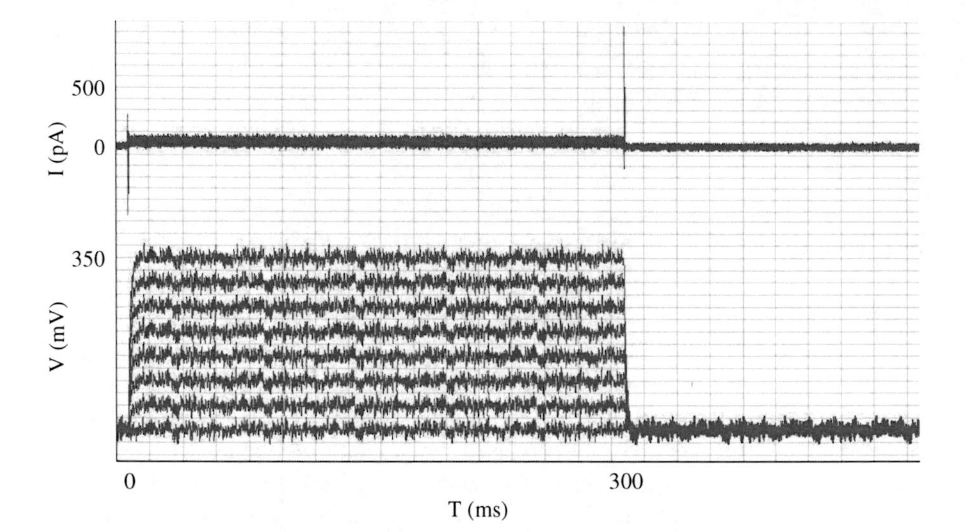

Fig. 5.8 Voltage clamp recordings showing changes in current performed by a piranha solution-treated pipette. The voltage step length is 30 ms, the increment is 50 mV per step. The application of a 350 mV pulse resulted in a recorded current of approximately 80 pA and a calculated seal resistance of 4.3 GΩ

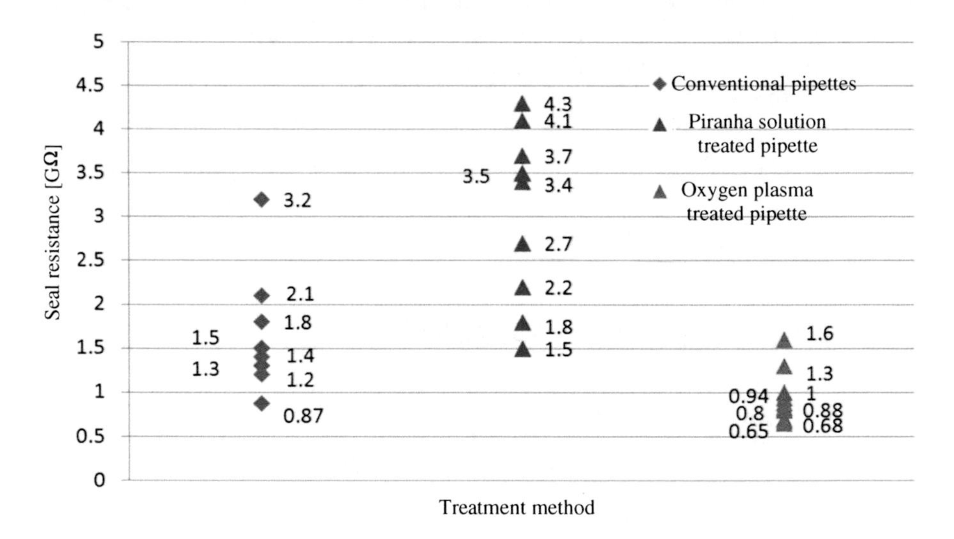

Fig. 5.9 Seal values for conventional, piranha solution and oxygen plasma-treated pipettes

5.4 Discussion

The results show that seal values for piranha solution-treated pipettes are higher than for conventional pipettes. Piranha solution treatment affects gigaseal formation in two ways: by increasing the chance for making hydrogen bonds and by cleaning the surface. The better performance of piranha solution-treated pipettes may be the result of these factors or a combination of both. As discussed in Sect. 3.3, four kinds of forces are presented in seal formation: ionic bonds, hydrogen bonds, salt bridges and van der Waals forces. Of these four, the last three play the more important role. Results from piranha solution-treated pipettes show the importance of hydrogen bonds in gigaseal formation. Piranha solution is a strong oxidizer and adds more hydroxyl groups to the surface of the glass, thus increasing the chance of forming hydrogen bonds between glass and membrane; therefore, a stronger seal is expected. Piranha solution also cleans the surface of the glass of any organic material. Cleanliness is a prerequisite for seal formation. These two improvements result in a very high probability of gigaseal formation: about 80 % of efforts led to gigaseal formation. The treatment is also easy in practice, which is highly desirable in electrophysiology labs; however, users need to be familiar with the safety procedures involved in using piranha solution.

Oxygen plasma treatment also cleans the surface and makes it more hydrophilic. However, oxygen plasma-treated pipettes did not form seals as good as piranha solution-treated or conventional pipettes. This might be mainly due to oxygen plasma not being an effective way of treating micropipettes. Firstly, due to the dimensions of the chamber, micropipettes should be placed horizontally. This configuration significantly limits the access of plasma to the inside of the pipette (Fig. 5.10a). Secondly, the inner wall of micropipettes should be treated for the first 100 μm from the pipette tip, as in seal formation the membrane goes from 5 to 100 microns into the pipette. Thirdly, the tip size is very small (1–2 μ) and as a result plasma cannot reach the desired area effectively. Fourthly, background gases present in the inducting coupled plasma etching chamber may contaminate the tip [13], thereby preventing seal formation. It should also be emphasized that oxygen plasma treatment has been widely used in planar patch clamp systems for treating the patching site [2, 4, 5]. The higher seal values after treatment show the importance of hydrophilicity of the patching site and the effectiveness of plasma treatment. In treating planar patch clamp systems, charged species can easily reach to the patching site because it is normally on a flat surface; therefore the site is effectively treated (Fig. 5.10b).

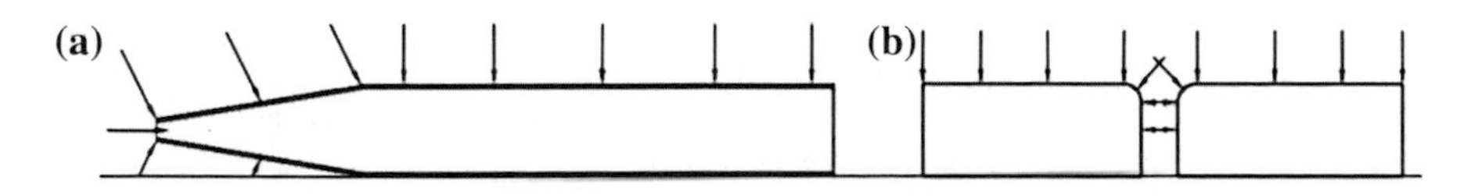

Fig. 5.10 Oxygen plasma treatment of **a** micropipettes, and **b** planar patch clamp chips, plasma can effectively treat the patching site in planar patch clamp systems but the inner walls of pipettes are difficult to access

5.5 Summary

In this chapter the effect of hydrophilicity on gigaseal formation has been reported. Treatments with piranha solution and oxygen plasma are used to change the hydrophilicity of the patching site. Both the piranha solution and oxygen plasma treatment increase surface roughness slightly, which can be neglected. Results show that piranha solution-treated pipettes form better seals with higher resistance values. These results can be understood by the fact that piranha solution is a strong oxidizer and adds more hydroxyl groups to the surface of the glass. Therefore more hydrogen bonds can be made between glass and membrane and a stronger seal is obtained. Piranha solution also cleans the surface of any organic material and cleanliness of patching area is a crucial requirement for seal formation [1]. Another advantage of piranha solution treatment is that the probability of gigaseal formation is very high. About 80 % of efforts using this method led to gigaseal formation, whereas oxygen plasma-treated pipettes failed to form high-resistance seals. This is mainly because oxygen plasma treatment is not an effective way to treat glass micropipettes. However, treatment of planar patch clamping chips with oxygen plasma, where there is good access of plasma to the patching site, has improved seal formation.

References

1. Hamill OP et al (1981) Improved patch-clamp techniques for high-resolution current recording from cells and cell-free membrane patches. Eur J Physiol 391:85–100
2. Klemic K, Klemic J, Sigworth F (2005) An air-molding technique for fabricating PDMS planar patch-clamp Electrodes. Eur J Physiol 449:564–572
3. Stett A et al (2003) CYTOCENTERING: A novel technique enabling automated cell-by-cell patch clamping with the CYTOPATCHTM chip. Recept Channels 9:59–66
4. Picollet-D'hahan N et al (ed) (2003) Multi-patch: a chip-based ionchannel assay system for drug screening. ICMENS international conference on MEMS, NANO & smart systems. Alberta Canada, pp 251–254
5. Wilsona S et al (2007) Automated patch clamping systems design using novel materials. 4 M annual conference. Borovets
6. Ionescu-Zanetti C et al (2005) Mammalian electrophysiology on a microfluidic platform. In: Proceeding of the National Academy of Science of the United States of America (PNAS), vol 102, pp 9112–9117
7. Williams KR, Gupta K, Wasilik M (1996) Etch rates for micromachining processing. J Microelectromech Syst 5:4, 256–269
8. Ming Mao C, Jung Hua C (2010) Advances in selective wet etching for Nanoscale NiPt salicide fabrication. Jpn J Appl Phys 4:1–5
9. Ziegler KJ et al (2005) Cutting single-walled carbon nanotubes. Nanotechnology 16:539–544
10. Seu KJ et al (2007) Effect of surface treatment on diffusion and domain formation in supported lipid bilayers. Biophys J 92:2445–2450
11. Pasquariello D, Hjort K (2002) Plasma-assisted InP-to-Si low temperature Wafer bonding. IEEE J Sel Topics Quantum Electron 8:118–131

12. Seung Woo C et al (2002) The analysis of oxygen plasma pretreatment for improving anodic bonding. J Electrochem Soc 149:8–11
13. Toshiaki Y et al (2004) Improvement on hydrophilic and hydrophobic properties of glass surface treated by nonthermal plasma induced by silent corona discharge. Plasma Chem Plasma Process 24:1–12

Chapter 6
Effect of Tip Size on Gigaseal Formation

Tip size is perhaps the easiest controllable factor which affects gigaseal formation. It is generally known that the smaller the tip size, the easier it is to achieve a gigaseal. Although tip size has been mentioned in the literature as an important factor in seal formation and has often been used in planar patch clamping as a tool to increase the probability of gigaseal formation [1–5], the reasons of the observation remain unclear. This chapter reports the research into the effect of tip size on gigaseal formation. The approach adopted in the research includes, first, investigation on how the tip size affects the patch clamping, then a study on the surface properties of pipettes with different tip sizes and compare the seal resistances obtained by them, and finally verification of the conclusions using experiments.

6.1 Effect of Tip Size in Patch Clamping

The resistance at the micropipette tip plays an important role in patch clamping since it constitutes the most of the pipette resistance. The other part of pipette resistance is due to the shank of the pipette and is called shank resistance. The simplified equivalent circuit for the cell attached is shown in Fig. 6.1. According to Kirchhoff's voltage law, the greatest voltage drop in a series circuit will be over the highest resistance. It means that the highest resistance in a series circuit determines the current flow. Therefore if the patch resistance (R_{patch}) is high compared with the resistance of the rest of the cell ($R_{membrane}$) and the pipette resistance ($R_{pipette}$), then the circuit effectively monitors current flow through the patch and any ion channels in it.

Pipette resistance can be obtained theoretically by knowing the geometry of patch pipettes. The tip shape is approximately conical, with an angle φ of 8–12°. When the pipette is modelled as having an approximately cylindrical shank and a conical tip, the total resistance of the pipette is given by the sum of the tip and the shank resistances [3].

$$R = \frac{\rho l}{\pi r_s^2} + \frac{\rho \cot\left(\frac{\varphi}{2}\right)}{\pi}\left(\frac{1}{r_t} - \frac{1}{r_s}\right) \tag{6.1}$$

M. Malboubi and K. Jiang, *Gigaseal Formation in Patch Clamping*, SpringerBriefs in Applied Sciences and Technology,
DOI: 10.1007/978-3-642-39128-6_6, © The Author(s) 2014

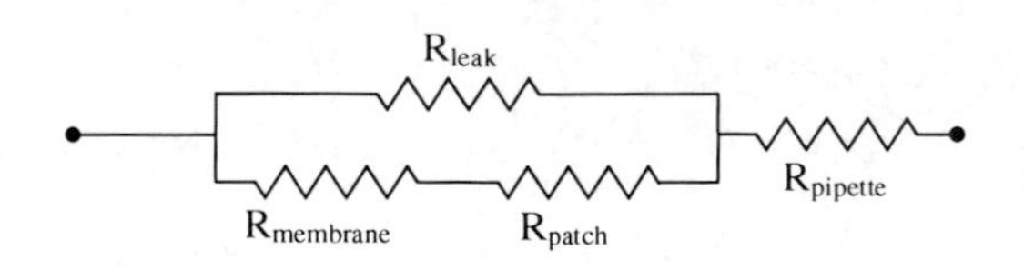

Fig. 6.1 Simplified equivalent circuit for the cell-attached patch configuration

where:

ρ is specific resistivity (Ohm.cm)
l is pipette shank length (cm)
r_s is the shank radius (μm)
r_t is the tip radius (μm)
φ is the cone angle.

Since the radius of the cylindrical shank (r_s) is much larger (>50 μm) than that of the radius of the tip opening (r_t), the resistance of the tip dominates. Most of the resistance of a patch pipette resides at or very near its tip.

As was discussed earlier, lower pipette resistance results in more accurate measurement of membrane activity. However the tip size cannot be increased greatly, since it lowers the probability of seal formation. One solution is to minimize shank resistance as much as possible by selecting correct pulling parameters in conventional patch clamping or suitable chip design in planar patch clamping devices.

6.2 Effect of Tip Size on Gigaseal Formation

Tip size affects gigaseal formation. The smaller the tip is, the higher the seal resistance. A cell membrane has contact with the pipette in two areas: the pipette tip and the pipette inner wall (Fig. 6.2).

In patch clamping, the membrane can be sucked into the pipette from 5 to 100 μm [6–9]. This implies that as long as the membrane and pipette are in close contact their length of opposition is of secondary importance. This suggests that the membrane can get closer to the pipette surface in the case of smaller pipettes. Measuring surface properties of pipettes with different sizes can help prove this hypothesis.

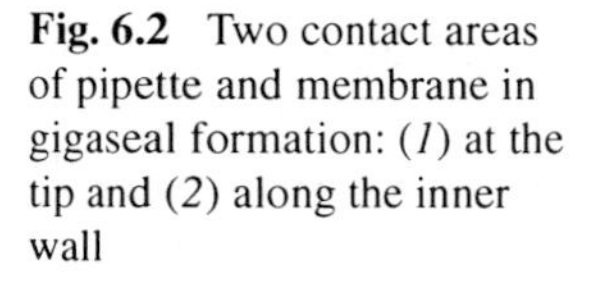

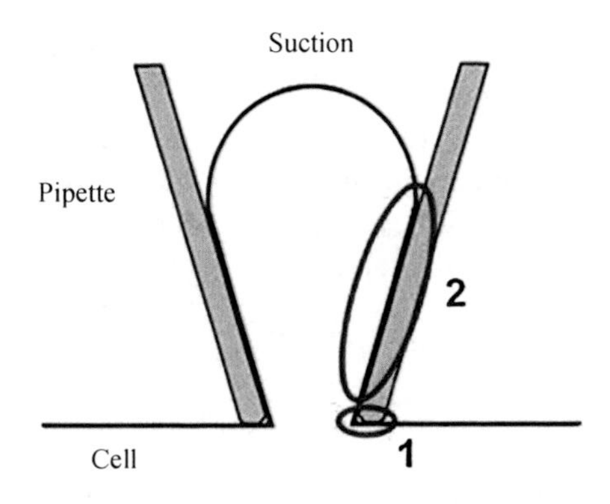

Fig. 6.2 Two contact areas of pipette and membrane in gigaseal formation: (*1*) at the tip and (*2*) along the inner wall

6.3 Measuring Surface Properties of Pipettes

The SEM stereoscopic technique is used to determine three-dimensional surface structures of pipettes.

6.3.1 3D Reconstruction of the Tip

To enhance the quality of images to the level required by the SEM stereoscopic technique, glass micropipettes are coated with a less than 5 nm-thick layer of platinum. Figure 6.3 shows images of pipettes before and after coating. The coating

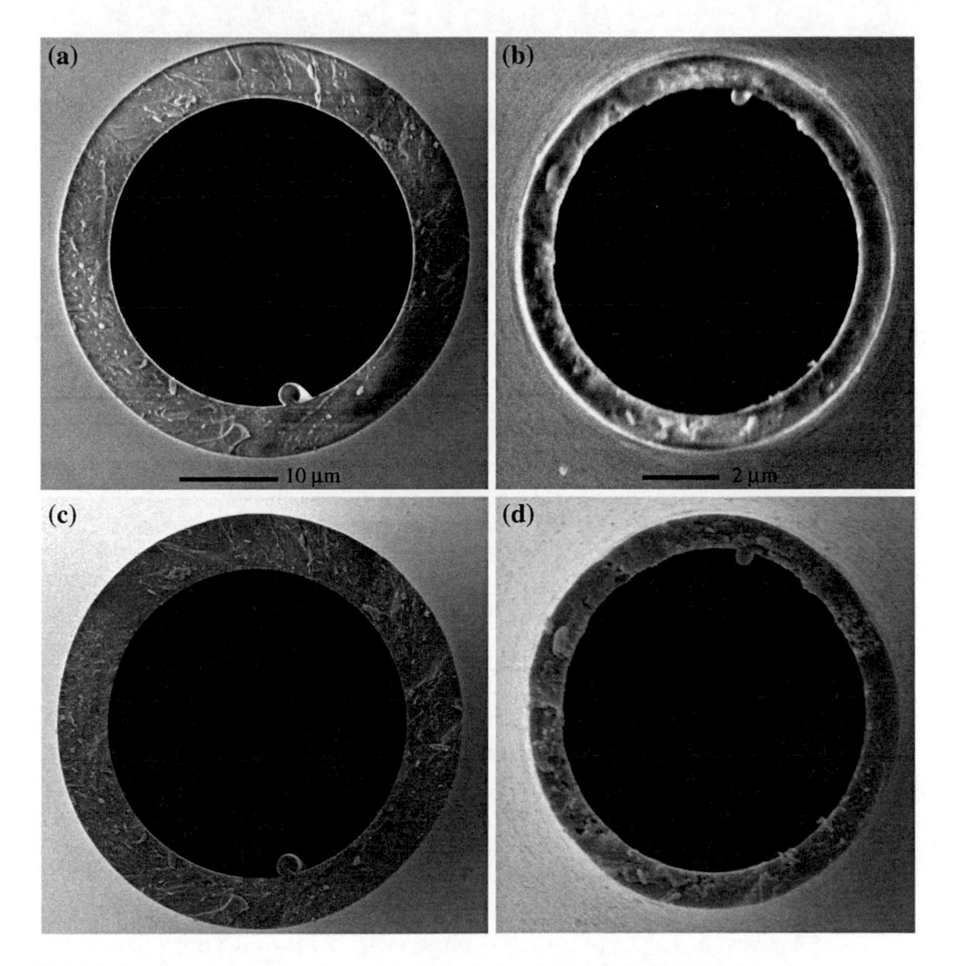

Fig. 6.3 Images of pipette tips before (**a**, **b**) and after (**c**, **d**) coating with platinum

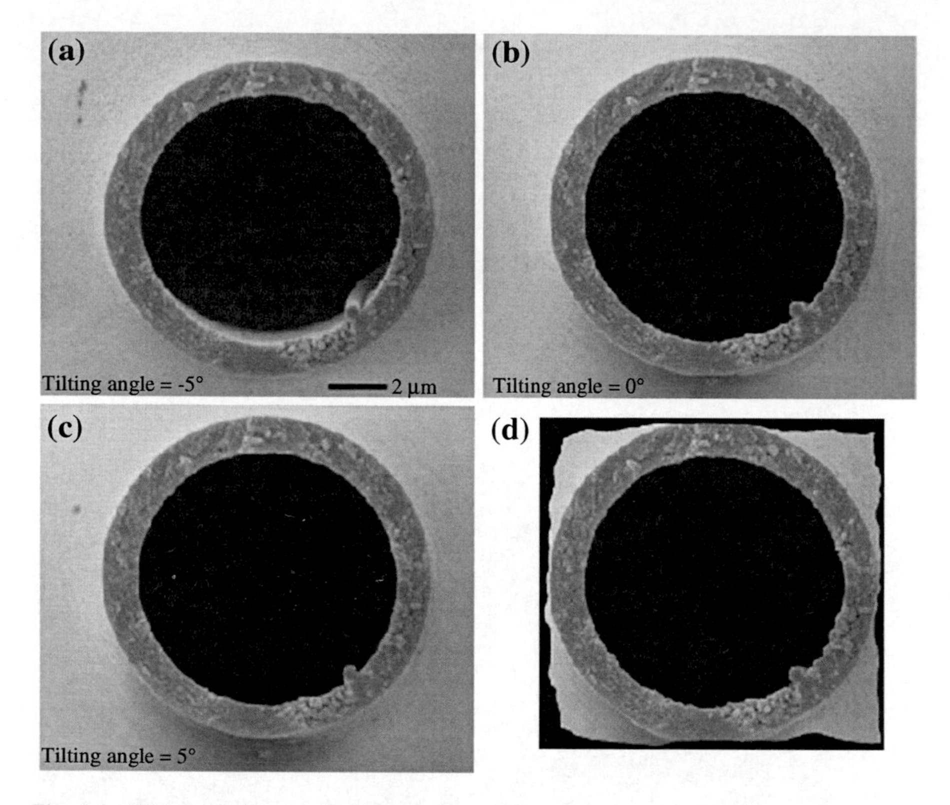

Fig. 6.4 SEM stereo images of pipette A (D_t = 8.7 μm): **a** west, **b** middle, **c** east images and **d** the digital elevation model created by MeX™

improved significantly the quality of the SEM images. The coating thickness is well below the resolution of the images used in 3D reconstruction. Therefore surface features are not affected by the coating process.

Two pipettes with different tip sizes were chosen for 3D reconstruction:

- Pipette A with a tip size of 8.7 μm and
- Pipette B with a tip size of 2.3 μm.

To reconstruct the pipettes' tips three high-resolution SEM images were obtained from different perspectives. The tilting angle between each pair of stereo images was 5 degrees. The SEM machine used for 3D reconstructions is Strata DB 235 from FEI. MeX™ software (version 5.1) was used for analyses and 3D reconstruction of the pipette surface [10]. Figure 6.4 shows the west, the middle and the east SEM images of pipette A and also the digital elevation model created by MeX™.

Figure 6.5 shows stereo images and the digital elevation model of pipette B.

Table 6.1 gives the values of tip sizes, tilting angles, magnifications, lateral resolutions and vertical resolutions of the two 3D reconstructed pipettes.

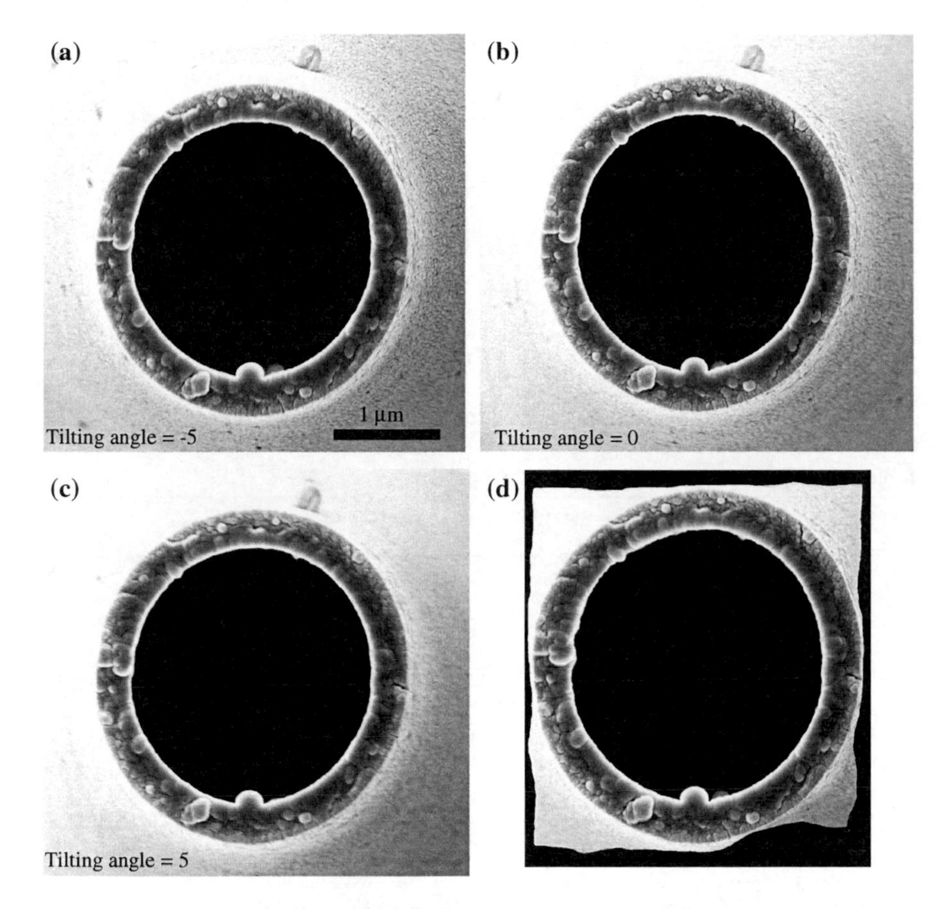

Fig. 6.5 SEM stereo images of pipette B (D_t = 2.3 μm): **a** west, **b** middle, **c** east images and **d** the digital elevation model created by MeX™

Table 6.1 Reconstruction information of pipettes A and B

Pipette	Tip size (μm)	Tilting angle (left to right)	Magnification	Lateral resolution (nm)	Vertical resolution (nm)
Pipette A	8.7	10	20000	14.7	42.1
Pipette B	2.3	10	65000	4.6	13.2

Surface properties of pipettes computed from the digital elevation model are presented in Table 6.2. The results show that pipette A is rougher than pipette B. Pipette A (D_t = 8.7 μm) has an average surface roughness of 40.8 nm and a maximum peak to valley distance of 585.9 nm. The average surface roughness of pipette B (D_t = 2.3 μm) is 17.3 nm and its maximum peak to valley distance is 204.8 nm. Table 6.3 shows the bearing area curve parameters of the pipettes. These parameters provide useful information about the peak, core and valley volumes and fluid retention ability of the surface [11] which will be used to explain the sources of leakage in seal formation later in this chapter.

Table 6.2 Tip surface properties of pipettes A and B

Name	Value (pipette A) $D_t = 8.7\ \mu m$	Value (pipette B) $D_t = 2.3\ \mu m$	Description
S_a	40.8 nm	17.3 nm	Average height of selected area
S_q	54.5 nm	22.6 nm	Root-mean-square height of selected area
S_p	258.7 nm	92.0 nm	Maximum peak height of selected area
S_v	327.2 nm	112.8 nm	Maximum valley depth of selected area
S_z	586 nm	204.8 nm	Maximum height of selected area
S_{10z}	438.7 nm	168.2 nm	Ten point height of selected area
S_{sk}	−0.1515	−0.2774	Skewness of selected area
S_{ku}	4.5606	3.9823	Kurtosis of selected area
S_{dq}	0.7303	0.8654	Root mean square gradient
S_{dr}	27.005 %	34.866 %	Developed interfacial area ratio

Table 6.3 Values of the bearing area curve of pipettes A and B

Name	Value (pipette A) $D_t = 8.7\ \mu m$	Value (pipette B) $D_t = 2.3\ \mu m$	Description
S_k	358.9 nm	116.9 nm	Core roughness depth, height of the core material
S_{pk}	129.1 nm	67.1 nm	Reduced peak height, mean height of the peaks above the core material
S_{vk}	236.0 nm	47.3 nm	Reduced valley height, mean depth of the valleys below the core material
S_{mr1}	12.4 %	10.18 %	Peak material component, the fraction of the surface which consists of peaks above the core material
S_{mr2}	90.74 %	89.23 %	Peak material component, the fraction of the surface which will carry the load
V_{mp}	0.0059 ml/m^2	0.0033 ml/m^2	Peak material volume of the topographic surface (ml/m^2)
V_{mc}	0.1225 ml/m^2	0.0411 ml/m^2	Core material volume of the topographic surface (ml/m^2)
V_{vc}	0.1816 ml/m^2	0.0563 ml/m^2	Core void volume of the surface (ml/m^2)
V_{vv}	0.021 ml/m^2	0.0056 ml/m^2	Valley void volume of the surface (ml/m^2)
V_{vc}/V_{mc}	1.4817	1.37	Ratio of V_{vc} parameter to V_{mc} parameter

6.3.2 3D Reconstruction of the Inner Wall

So far it has been shown that tips of micropipettes are rough and a larger pipette has a higher average surface roughness. It has been thought that heating and pulling of glass micropipettes produces smooth surfaces [12], and therefore one might expect that inner walls of pipettes to be smooth. To measure surface properties of the inner walls of pipettes, focused ion beam milling and SEM stereoscopic technique were used. Two pipettes with different sizes were chosen:

- Pipette C with a tip size of 13.2 μm and
- Pipette D with a tip size of 8.9 μm

The pipettes were split and cut open using focused ion beam milling to create access to the inner walls. The imaging direction was perpendicular to the cutting plane, avoiding redeposition of sputtered materials from the Focused Ion Beam (FIB) cutting onto the area. After cutting, the pipettes were turned 90° by means of a previously fabricated holder. Three SEM images were taken from the inside wall and 3D structures of the inner wall were obtained using MeX software. Figure 6.6 shows pipette C (D_t = 13.2 μm) before and after FIB milling. West, middle and east SEM images of the inner wall of this pipette and its digital elevation model created by MeX are shown in Fig. 6.7.

Figure 6.8 shows pipette D (D_t = 8.9 μm) before and after FIB milling. West, middle and east SEM images of the inner wall of this pipette and its digital elevation model created by MeX are shown in Fig. 6.9.

Table 6.4 gives the values of tip sizes, tilting angles, magnifications, lateral resolutions and vertical resolutions of the two 3D reconstructed pipettes.

Surface properties of the pipettes' inner wall surfaces computed from the digital elevation model are presented in Tables 6.5 and 6.6. The results show that pipette C (D_t = 13.2 μm) has an higher average surface roughness, of 30.2 nm. The average surface roughness of pipette D (D_t = 8.9 μm) is 24.1 nm.

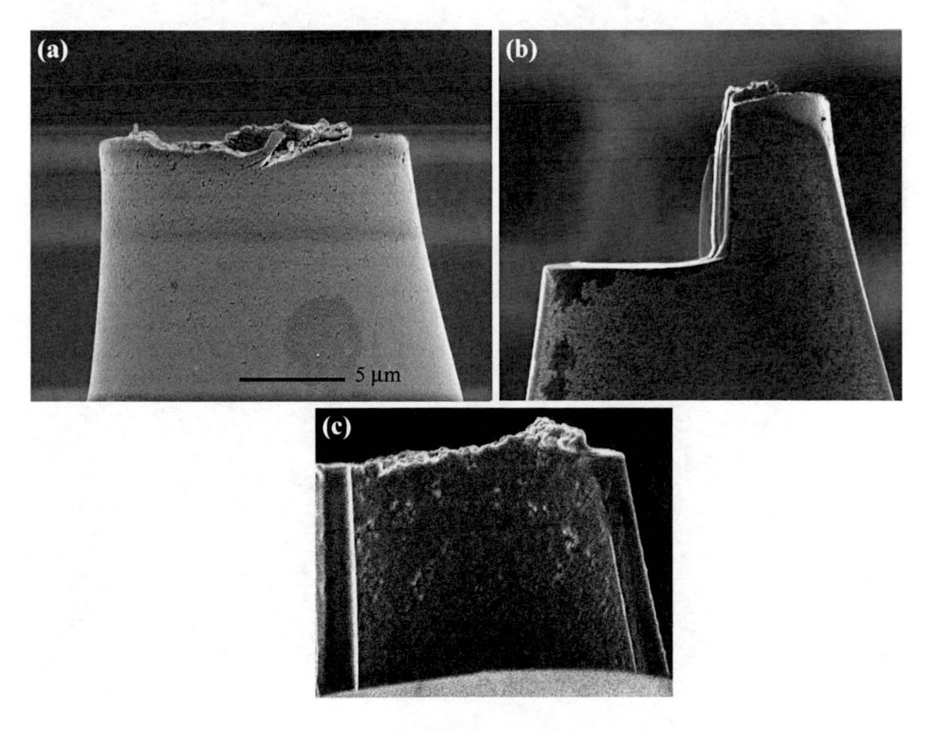

Fig. 6.6 Pipette C (D_t = 13.2 μm) *before* and *after* focused ion beam milling

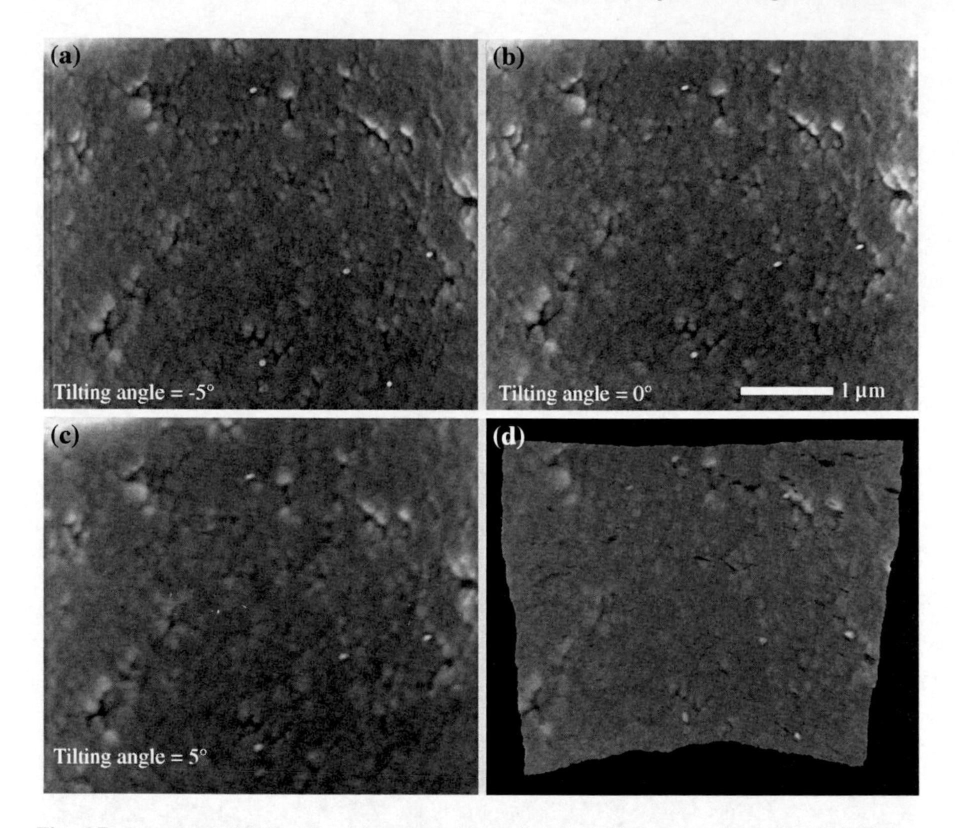

Fig. 6.7 Inner wall SEM stereo images of pipette C (D_t = 13.2 μm): **a** west, **b** middle, **c** east images and **d** the digital elevation model created by MeX™ [14]

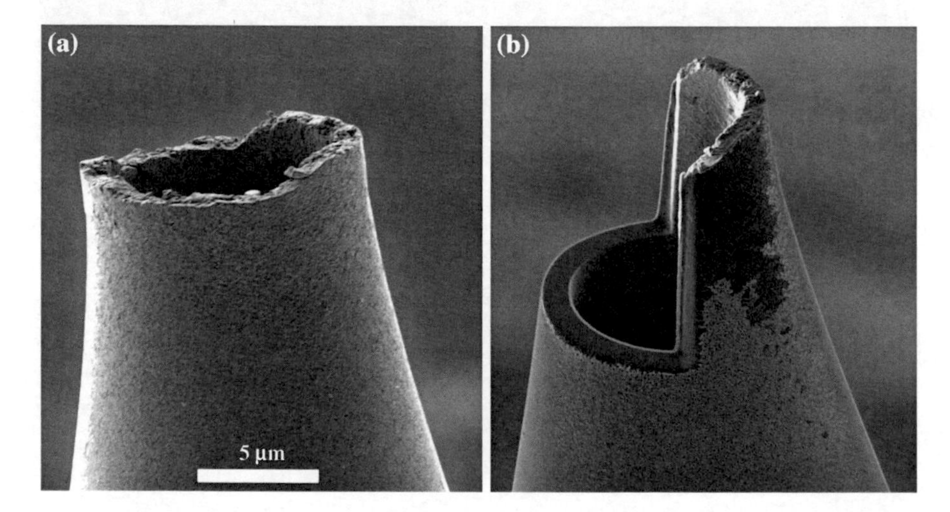

Fig. 6.8 Pipette D (D_t = 8.9 μm) *before* and *after* FIB milling [15]

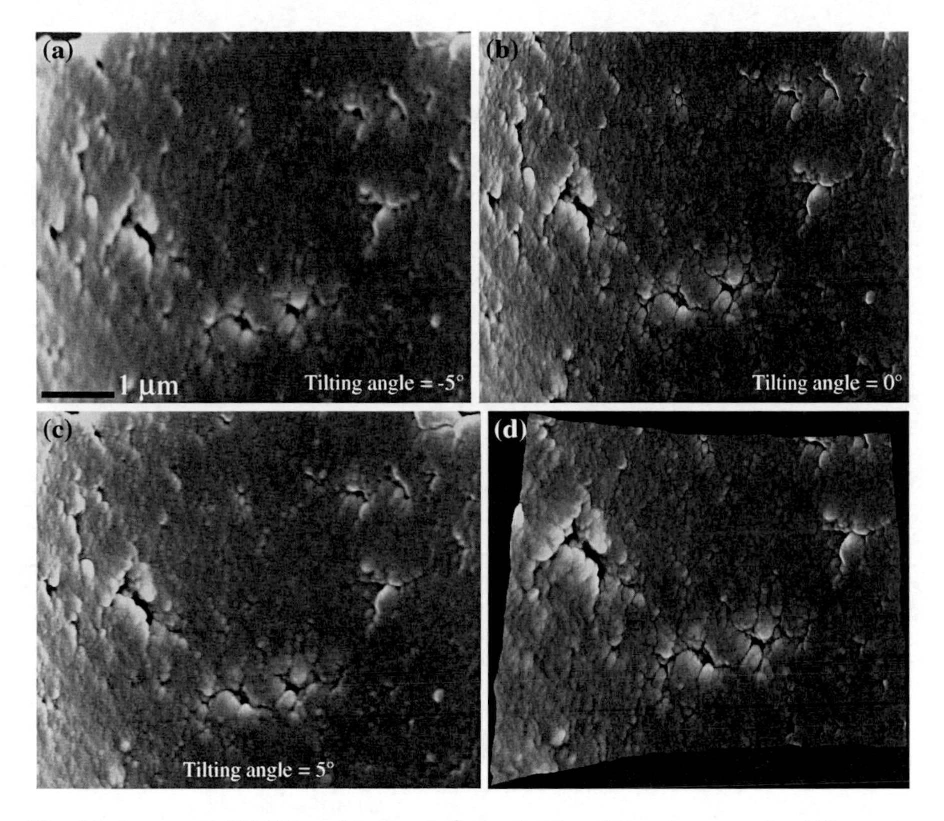

Fig. 6.9 Inner wall SEM stereo images of pipette D (D_t = 8.9 μm): **a** west, **b** middle, **c** east images and **d** the digital elevation model created by MeX™ [15]

Table 6.4 Reconstruction information of pipettes C and D

Pipette	Tip size (μm)	Tilting angle (left to right)	Magnification	Lateral resolution (nm)	Vertical resolution (nm)
Pipette C	13.2	10	25000	11.7	33.7
Pipette D	8.9	10	50000	5.88	16.8

6.4 Patch Clamp Experiments

Patch clamp experiments were carried out on HEK (Human Embryonic Kidney) cells using two kinds of pipette:

- Conventional pipettes with tip size of 1.1 μm and resistance of 6.5 MΩ
- Large pipettes with tip size of 3.5 μm and resistance of 1.8 MΩ.

Cell culture and set up for patch clamp experiments were the same as discussed in Chap. 4. Ten measurements were made for each type of pipette and seal values were recorded. Figures 6.10 and 6.11 show the seal quality for the conventional and larger pipettes respectively. Figure 6.12 shows the seal values for the

Table 6.5 Inner wall surface properties of pipettes C and D

Name	Value (pipette C) $D_t = 13.2\ \mu m$	Value (pipette D) $D_t = 8.9\ \mu m$	Description
S_a	30.2 nm	24.1 nm	Average height of selected area
S_q	37.9 nm	30.6 nm	Root-mean-square height of selected area
S_p	156.9 nm	163.0 nm	Maximum peak height of selected area
S_v	149.1 nm	145.9 nm	Maximum valley depth of selected area
S_z	351 nm	316 nm	Maximum height of selected area
S_{10z}	272.2 nm	249.0 nm	Ten point height of selected area
S_{sk}	0.0342	0.0696	Skewness of selected area
S_{ku}	2.9718	3.3903	Kurtosis of selected area
S_{dq}	0.6154	0.8312	Root mean square gradient
S_{dr}	17.827 %	30.966	Developed interfacial area ratio

Table 6.6 Values of the bearing area curve of pipettes C and D

Name	Value (pipette C) $D_t = 13.2\ \mu m$	Value (pipette D) $D_t = 8.9\ \mu m$	Description
S_k	458.8 nm	340.9 nm	Core roughness depth, height of the core material
S_{pk}	225.4 nm	174.2 nm	Reduced peak height, mean height of the peaks above the core material
S_{vk}	122.4 nm	47.9 nm	Reduced valley height, mean depth of the valleys below the core material
S_{mr1}	19.52 %	17.36 %	Peak material component, the fraction of the surface which consists of peaks above the core material
S_{mr2}	93.79 %	96.72 %	Peak material component, the fraction of the surface which will carry the load
V_{mp}	0.0079 ml/m^2	0.0062 ml/m^2	Peak material volume of the topographic surface (ml/m^2)
V_{mc}	0.1741 ml/m^2	0.1306 ml/m^2	Core material volume of the topographic surface (ml/m^2)
V_{vc}	0.2841 ml/m^2	0.2115 ml/m^2	Core void volume of the surface (ml/m^2)
V_{vv}	0.0155 ml/m^2	0.0073 ml/m^2	Valley void volume of the surface (ml/m^2)
V_{vc}/V_{mc}	1.6326	1.6194	Ratio of V_{vc} parameter to V_{mc} parameter

two kinds of pipette. The mean value of seal resistance for conventional pipettes is 1.6 GΩ with the standard deviation of 0.6 GΩ. The mean value of seal resistances for larger pipettes is 0.4 GΩ with a standard deviation of 0.2 GΩ.

6.5 Discussion

Patch clamping results clearly show that conventional pipettes ($D_t = 1.1\ \mu m$) make better seals. The average seal value is 1.6 GΩ for conventional pipettes, with a standard deviation of 0.6 GΩ, and is significantly higher than the average

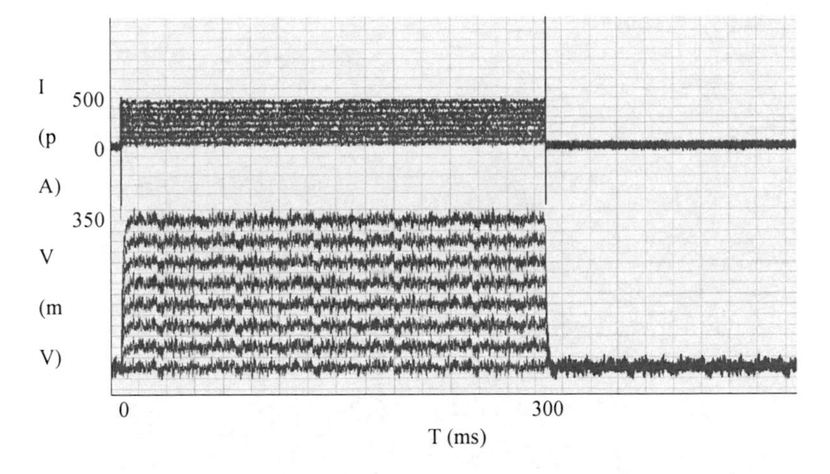

Fig. 6.10 Voltage clamp recordings showing changes in current performed by a larger pipette (tip size = 3.5). The voltage step length is 30 ms, and the increment is 50 mV per step. The application of a 350 mV pulse resulted in a recorded current of approximately 480 pA and a calculated seal resistance of approximately 0.7 GΩ

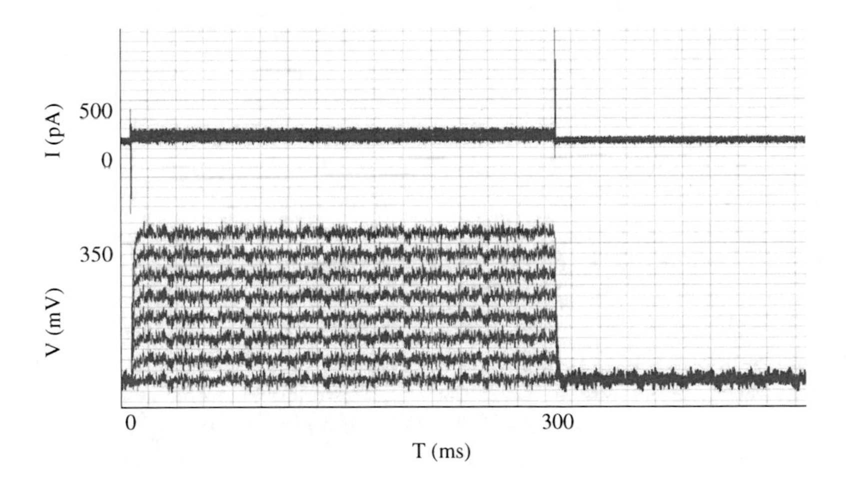

Fig. 6.11 Voltage clamp recordings showing changes in current performed by a conventional pipette (tip size = 1.1). The voltage step length is 30 ms, and the increment is 50 mV per step. The application of a 350 mV pulse resulted in a recorded current of approximately 110 pA and a calculated seal resistance of approximately 3.2 GΩ

seal value of 0.4 GΩ for larger pipettes with a diameter of 3.5 μm. This can be explained by comparing the roughness parameters of pipettes with different sized openings. Tables 6.2 and 6.5 show the surface roughness parameters for pipettes with different tip sizes. The fact that both the pipette tips and the pipette inner walls are rough may help towards a better understanding of the

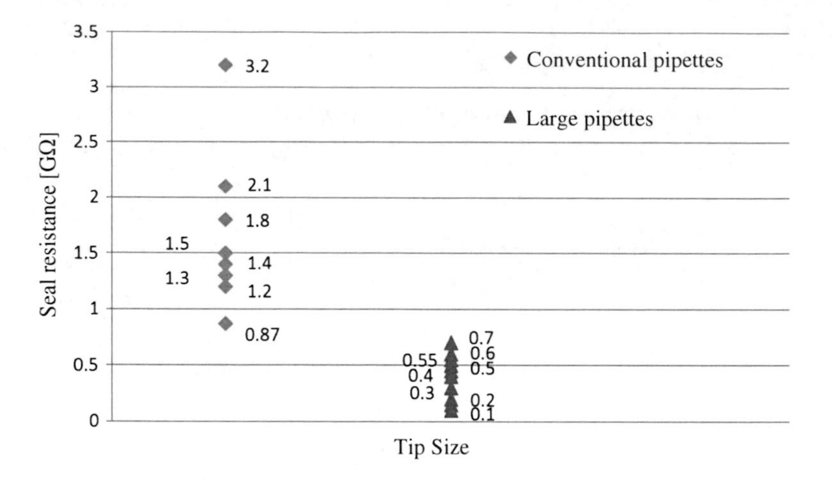

Fig. 6.12 Seal values for conventional and larger pipettes

mechanism of gigaseal formation. The larger pipettes (pipettes A and C) have a higher average surface roughness (S_a), a higher maximum peak to valley distance and a lower developed interfacial ratio (S_{dr}). From Chap. 3, it is known that maximum peak to valley distance determines the height of the channel connecting the inside of the pipette to the outside. Higher maximum peak to valley distances for pipettes A and C show that the inside of these pipettes is connected to the outside by larger channels facilitating the leakage of ions. The developed interfacial area ratio (S_{dr}) also changes significantly for pipettes having different sizes. S_{dr} is expressed as the percentage of additional surface area contributed by the texture as compared to an ideal plane [11]. Higher S_{dr} means that the surface is closer to a flat surface. It has been shown that higher S_{dr} promotes cell adhesion significantly [13] by allowing the membrane to get closer to the glass surface. As a result more bonds can be made between cell surface and glass wall. The fact that pipettes B and D have notably higher S_{dr} at the tip and at the pipette inner wall surface means that a higher percentage of the pipette surface contributes in glass-membrane interactions. This increases the number of membrane proteins sticking to the pipette inner wall and improves the seal. Tables 6.3 and 6.6 show that valley void volume (V_{vv}) is considerably higher for pipettes A and C (i.e. the larger pipettes). This indicates that these pipettes have a greater fluid retention ability. The ratio of V_{vc}/V_{mc} is also larger for them, which means that there are more voids present compared to pipettes B and D (i.e. the smaller pipettes). During patch clamp experiments valleys and voids are filled with conductive media, facilitating ion escape, thereby increasing the leakage current and compromising the seal. The results suggest that as long as the membrane and pipette surface are close enough, the length of the contact is of secondary importance.

6.6 Summary

In this chapter the effect of tip size on gigaseal formation has been studied. The surface roughness parameters of pipettes of different sizes were measured using a SEM stereoscopic technique. In order to have access to the inner wall of the pipette, the pipettes' heads were split and cut open using FIB milling. It was found that the larger pipettes have higher average surface roughness, higher maximum peak to valley distance, higher valley void volume and lower developed interfacial area ratio. These findings explain the higher leakage current and lower seal resistance in the case of larger pipettes. The results are in good agreement with the practical knowledge in patch clamping, that the smaller pipette makes the better seal.

References

1. Klemic K, Klemic J, Sigworth F (2005) An air-molding technique for fabricating PDMS planar patch-clamp Electrodes. Eur J Physiol 449:564–572
2. Kusterer J et al (2005) A diamond-on-silicon patch-clamp-system. Diam Relat Mater 14:2139–2142
3. Sakmann B, Neher E (2009) Single-channel recording, 2nd edn. Springer, New York. doi:978-1-4419-1230-5
4. Petrov AG (2001) Flexoelectricity of model and living membranes. Biochim Biophys Acta 1561:1–25
5. Petrov AG (2006) Electricity and mechanics of biomembrane systems: flexoelectricity in living membranes. Anal Chim Acta 568:70–83
6. Nagarah JM et al (2010) Batch fabrication of high-performance planar patch-clamp devices in quartz. Adv Mater 22:4622–4627
7. Priel A et al (2007) Ionic requirements for membrane-glass adhesion and gigaseal formation in patch-clamp recording. Biophys J 92:3893–3900
8. Ruknudin A, Song MJ, Sachs E (1991) The ultrastructure of patch-clamped membranes: a study using high voltage electron microscopy. J Cell Biol 112:125–134, 1
9. Sokabe M, Sachs F, Jing Z (1991) Quantitative video microscopy of patch clamped membranes stress, strain, capacitance, and stretch channel activation. Biophysl J 59:722–728
10. MeXTM software. Alicona Imaging GmbH, Graz, Austria
11. Stout KJ, Blunt L (2000) Three-dimensional surface topography. Penton Press, London, 1857180267
12. Lepple-Wienhues A et al (2003) Flip the tip: an automated, high quality, cost-effective patch clamp screen. Recept Channels 9:13–17
13. Eliaz N et al (2009) The effect of surface treatment on the surface texture and contact angle of electrochemically deposited hydroxyapatite coating and on its interaction with bone-forming cells. Acta. Biomater 5:3178–3191
14. Malboubi M, Gu Y, Jiang K (2011) Characterization of surface properties of glass micropipettes using SEM stereoscopic technique. Microelectron Eng 88:2666–2670
15. Malboubi M, Gu Y, Jiang K (2011) Surface properties of glass micropipettes and their effect on biological studies. Nanoscale Res Lett 6:1–10, 401

Chapter 7
Study of Glass Micropipettes from Tip Formation to Characterization

In this chapter various aspects of glass micropipettes are studied, including mechanisms of tip formation, tip geometry, and effect of pulling parameters on surface roughness properties of glass micropipettes. The study is intended both to explain some sources of leakage in patch clamping and to provide useful information for fabricating pipettes with favoured properties. The study may also lead to better understanding of the mechanism of tip formation.

7.1 Measuring Surface Properties of Glass Tubes

Glass micropipettes are fabricated by the heating and pulling of commercially available glass tubes. Surface properties of glass tubes (BF150-86-10, Sutter Instruments) were measured using light interferometers. White light interferometers allow the rapid acquisition of three-dimensional topographical information in order to create accurate maps of surface architectures. The shape and phase of interferometric fringes created by the optical path differences caused by the sample surface features, when compared to a reference mirror, allows the measurement of topographic information as the sample is scanned vertically relative to the instrument lens. Interferometric measurements of micropipettes were performed using a MicroXAM2 interferometer (Omniscan, UK), operating using a white light source. Pipettes were imaged at a magnification of 100X. Scanning Probe Image Processor software (Image Metrology, Denmark) was employed for the analysis of the acquired images and for the obtaining of surface roughness parameters. Figure 7.1 a–c shows the images of the outer wall of the pipette. Tables 7.1 and 7.2 show roughness parameters of the outer and inner wall of glass tubes.

These measurements clearly show that both of the outer and inner surfaces of glass tubes are rough before pulling. The roughness could come from the manufacturing process of the glass tubes.

M. Malboubi and K. Jiang, *Gigaseal Formation in Patch Clamping*,
SpringerBriefs in Applied Sciences and Technology,
DOI: 10.1007/978-3-642-39128-6_7,

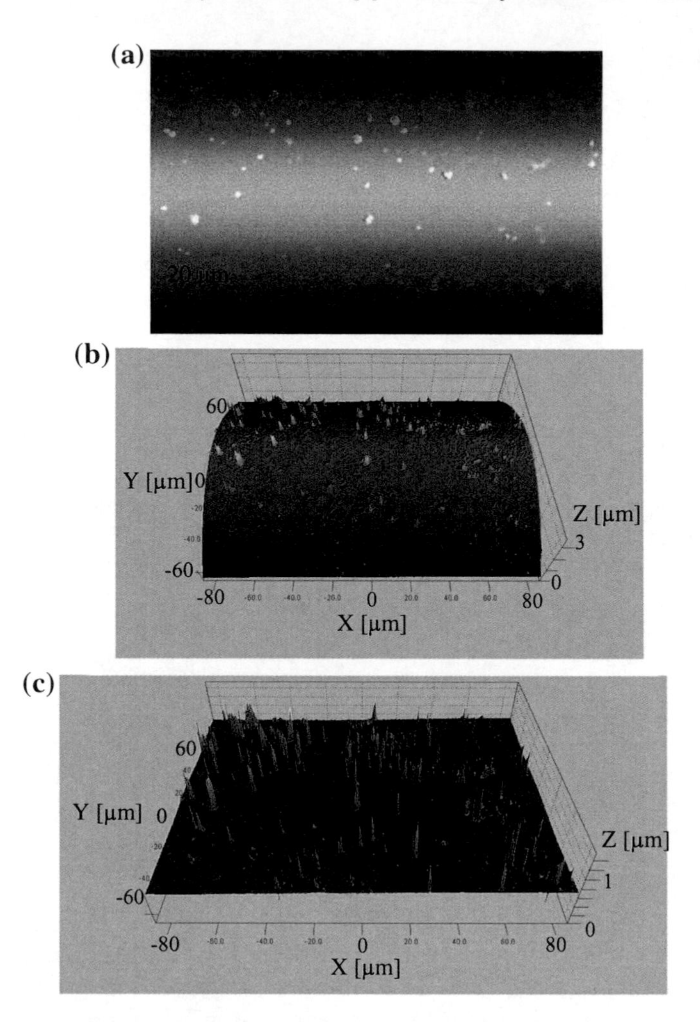

Fig. 7.1 Measurement of surface properties of glass tubes using light interferometry, **a**, **b** 2D and 3D representation of topology of pipette outer wall respectively, **c** representation of the pipette surface defects

7.2 Effect of Pulling Parameters on Pipette Tip Size and Surface Properties

The effect of pipette tip roughness and size on gigaseal formation has been studied in Chaps. 4 and 6 respectively. In this chapter, the effect of pulling parameters on these factors has been studied. The study provides means for controlling the size and roughness of micropipette tips, means which in addition to facilitating seal formation, can also be useful in many other applications. Glass micropipettes have frequently been used in applications such as: controlled delivery of liquids, genes

Table 7.1 Roughness parameters of the outer wall of glass tubes used in fabrication of glass micropipettes

Name	Value	Unit	Description
S_a	0.0197	μm	Average surface height of selected area
S_q	0.0431	μm	Root-mean-square height of selected area
S_{sk}	6.66		Skewness of selected area
S_{ku}	70.4		Kurtosis of selected area
S_y	1.54	μm	Largest peak to valley height
S_z	1.26	μm	Ten point height
S_{ds}	0.403	$1/\mu m^2$	Density of summits—number of summits of a unit sampling area
S_{sc}	0.0141	$1/\mu m$	Arithmetic mean summit curvature of the surface—average of the principal curvatures of the summits within the sampling area
S_{max}	1.54	μm	Maximum height of selected area
S_{2A}	22071	μm^2	Projected area
S_{3A}	22163	μm^2	Actual surface area

Table 7.2 Roughness parameters of the inner wall of glass tubes used in fabrication of glass micropipettes

Name	Value	Unit	Description
S_a	0.0188	μm	Average surface height of selected area
S_q	0.0294	μm	Root-mean-square height of selected area
S_{sk}	11.9		Skewness of selected area
S_{ku}	575		Kurtosis of selected area
S_y	2.80	μm	Largest peak to valley height
S_z	1.54	μm	Ten point height
S_{ds}	0.549	$1/\mu m^2$	Density of summits—number of summits of a unit sampling area
S_{sc}	0.0049	$1/\mu m$	Arithmetic mean summit curvature of the surface—average of the principal curvatures of the summits within the sampling area
S_{max}	2.80	μm	Maximum height of selected area
S_{2A}	22075	μm^2	Projected area
S_{3A}	22100	μm^2	Actual surface area

or sperm to the target [1–3], fertilization studies [4], intracellular measurements [5], voltage and current clamp studies [6]. Recent developments in micro-engineering and nanosciences have also led to new applications for micro-/nanopipettes, such as generating micro-droplets [7], single-molecule fluorescence tracking [1], creating nanoscale features by nanolithography and nanowriting methods [8], and nanosensing in scanning probe microscopy [9]. In many of these applications a smooth tip is preferred because it reduces the chance of tip contamination and damage to delicate biological samples [4]. Dozens of pipettes may be used by an individual in a single day. A small improvement in the condition of such pipettes may have a great influence on the outcomes. Despite the wide range of application, there have been few reports about numerical analysis on the effect of pulling parameters on the surface roughness properties of glass micropipettes.

7.2.1 Pulling Pipettes

The puller used in the experiments was a Flaming/Brown micropipette puller (Model P-97, Sutter Instruments, Novato, CA). The six parameters on this machine for controlling the shape and size of micropipettes are heat, pull, velocity, delay, time and pressure. These parameters are briefly introduced here, but full details of them can be found in the manufacturer's catalogue [10].

- HEAT (Range 0–999): HEAT controls the level of electrical current supplied to the filament. The units of heat are in milliamps. Useful changes in HEAT are 5 units or more to see an effect.
- VELOCITY (Range 0–255): The velocity of the glass carriage system is measured as the glass softens and begins to pull apart under a constant load. The velocity transducer is a patented approach [11] and this picks up on the velocity of the puller bars as the glass softens [12].
- PULL (Range 0–255): This parameter controls the force of the hard pull. The units of PULL determine the current to the pull solenoid. Useful changes in PULL strength are 10 units or more to see an effect.
- DELAY (Range 0–255): DELAY is a cooling mode which controls the delay time between when the heat turns off and when the hard pull is activated. One unit of DELAY represents ½ ms.
- TIME (Range 0–255): TIME is a cooling mode and controls the length of time the cooling air is active. One unit of TIME represents ½ ms.
- PRESSURE (Range 0–999): This control sets the pressure generated by the air compressor during the active cooling phase of the pull cycle. The units of pressure are in psi. Changes of less than 10 units will not be noticeable.

To investigate the effect of each parameter on the pipette's tip surface properties, one parameter was varied whereas the others were kept unchanged in every set of experiments. Delay and time are both cooling parameters. Time has quite a narrow working range, whereas delay provides a wider range of control; therefore the effect of delay is investigated. Table 7.3 shows values of the parameters used in the experiments. Glass micropipettes pulled from borosilicate glass tubes have an outer diameter of 1.5 mm and an inner diameter of 0.86 mm (BF150-86-10, Sutter Instruments). The filament of the puller machine was FB230B (2.0 mm square box filament, 3.0 mm wide, Sutter Instruments). Pulling pipettes continuously will make the chamber warm and gradually decrease the heating time for subsequent pipettes. For this reason the chamber was left for 5 min to cool down after pulling every five pipettes.

To test the reproducibility of the puller, 10 pipettes were pulled with a set of parameters and their tip sizes were measured by Scanning Electron Microscopy

Table 7.3 Pulling parameter values

Experiment	Heat	Velocity	Pull	Delay	Pressure
Effect of heat	595, 600 605	10	0	1	500
Effect of velocity	606	4, 8, 10, 12, 14	0	1	500
Effect of pull	606	10	0, 10, 30	1	500
Effect of delay	606	10	0	1, 20, 40	500
Effect of pressure	606	10	0	1	300, 350, 400, 450

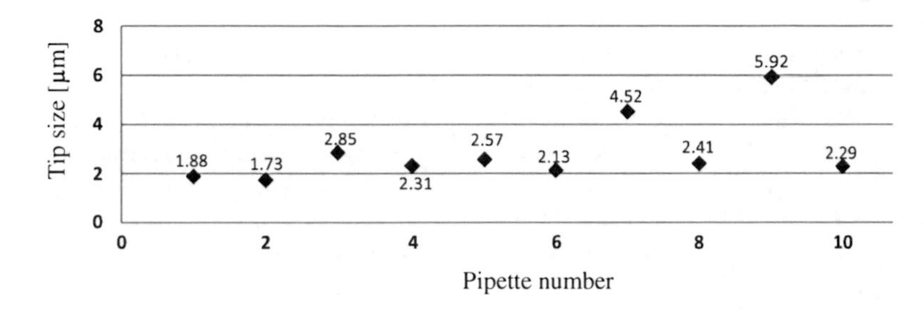

Fig. 7.2 Pipette pulling experiment records. Ten pipettes were pulled with the same pulling parameters and their tip sizes were measured using SEM

(SEM). Figure 7.2 is a summary of the statistics of the experiments. A few sudden variations in tip sizes are due mainly to the non-homogeneities in the composition and molecular structure of borosilicate glass [5]. In the experiments, pipettes with irregular sizes, far from the expected values, were not used for reconstruction.

7.2.2 3D Reconstruction of Pipette Tips

To determine the three-dimensional structure of pipette tips, the SEM stereoscopic technique was used in the investigation. Over 20 pipettes have been reconstructed. To capture high-quality SEM images which satisfy stereoscopic technique requirements, glass micropipettes were coated with a thin layer of platinum (<5 nm). The SEM machine used for 3D reconstruction was the Strata 235 SEM/FIB dual-beam system (FEI, Hillsboro, Oregon). The important factors in the SEM stereoscopic technique are magnification, tilting angle and resolution. Since the maximum pixel resolution of the machine is limited, different magnifications and tilting angles have been used to reconstruct every pipette's tip with maximum disparity and highest lateral and vertical resolution. Such a reconstruction could be expected to have an inaccuracy of less than 5 % [13]. Table 7.4 gives the values of tip size, tilting angle, magnification, lateral resolution and vertical resolution for three different-sized reconstructed pipettes. Surface properties of the largest and smallest

Table 7.4 Reconstruction information for three pipettes

Pipette number	Tip size (μm)	Tilting angle (left to right)	Magnification	Lateral resolution	Vertical resolution
1	34.5	10	5000	29 nm	41 nm
2	19.3	10	8000	18.1 nm	19.7 nm
3	3.7	10	50000	5.8 nm	8.2 nm

Table 7.5 The surface properties of the largest and smallest pipettes presented in Table 7.4

Name	Value ($D_t = 34.5$ μm)	Value ($D_t = 3.7$ μm)	Description
S_a	149.1 nm	30.8 nm	Average height of selected area
S_q	209.9 nm	42.0 nm	Root-mean-square height of selected area
S_p	1437.4 nm	304.1 nm	Maximum peak height of selected area
S_v	1409.3 nm	238.0 nm	Maximum valley depth of selected area
S_z	2846.7 nm	542.1 nm	Maximum height of selected area
S_{10z}	2304.3 nm	414.59 nm	Ten point height of selected area
S_{sk}	−0.2118	−0.514	Skewness of selected area
S_{ku}	7.0253	6.4208	Kurtosis of selected area
S_{dq}	0.5767	1.3246	Root mean square gradient
S_{dr}	13.129 %	79.532 %	Developed interfacial area ratio

pipette are shown in Table 7.5. Figure 7.3 shows the SEM stereo images and digital elevation model of one of the pipettes.

7.2.3 Effect of Pulling Parameters on Pipette Surface Properties

The effect of each parameter is studied by investigating at least three reconstructions. Tip size (D_t) and average surface roughness (S_a) of all pipettes were measured. Figures 7.4, 7.5, 7.6, 7.7, and 7.8 show correlations between pulling parameters and D_t and S_a.

As can be seen from Figs. 7.4, 7.5, 7.6, 7.7, and 7.8 velocity has the most significant effect. A small increase in velocity significantly decreases D_t and S_a. The effects of pull and heat are very similar and not as significant as the effect of velocity. Delay and pressure are factors to change the shank length of the pipettes while keeping the tip size unchanged [10]. Increasing delay and pressure will result in a shorter shank. Although these two factors do not change tip size significantly, it can be seen from Figs. 7.7 and 7.8 that the larger pipette has a higher surface roughness. From Figs. 7.4, 7.5, 7.6, 7.7, and 7.8 it can be understood that D_t and S_a have a direct correlation. Figure 7.9 is obtained by plotting D_t versus S_a for 21 pipettes pulled with different pulling parameters. It can be seen that that average surface roughness of a pipette is strongly related to tip size. D_t and S_a

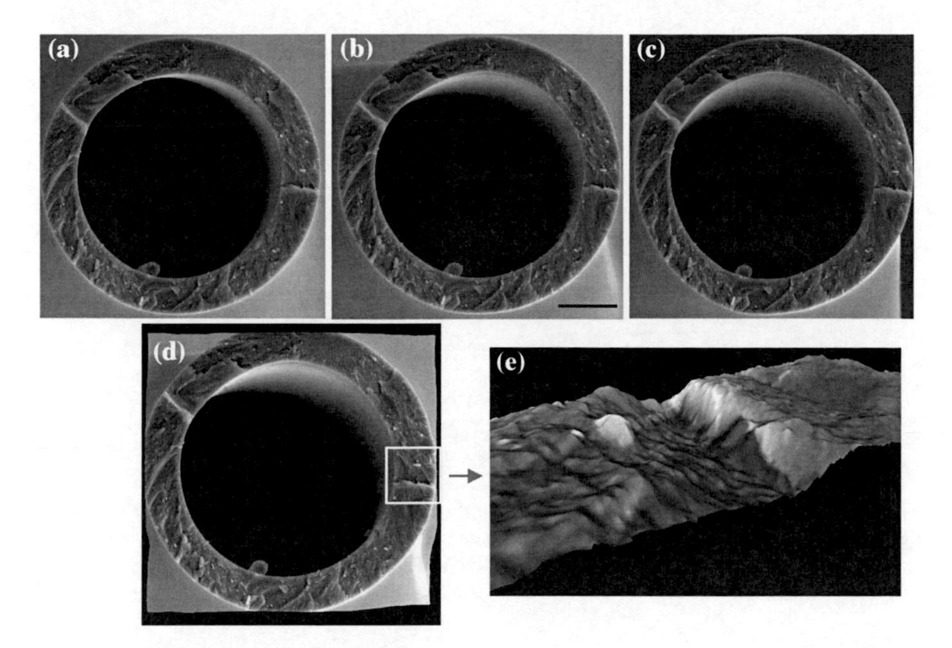

Fig. 7.3 SEM stereoscopic images captured from different angles: **a** −5 degrees, **b** 0 degree and **c** 5 degrees. **d** Digital elevation model created using MeX. **e** The exploded view of the selected area. The bar represents 10 μm [14]

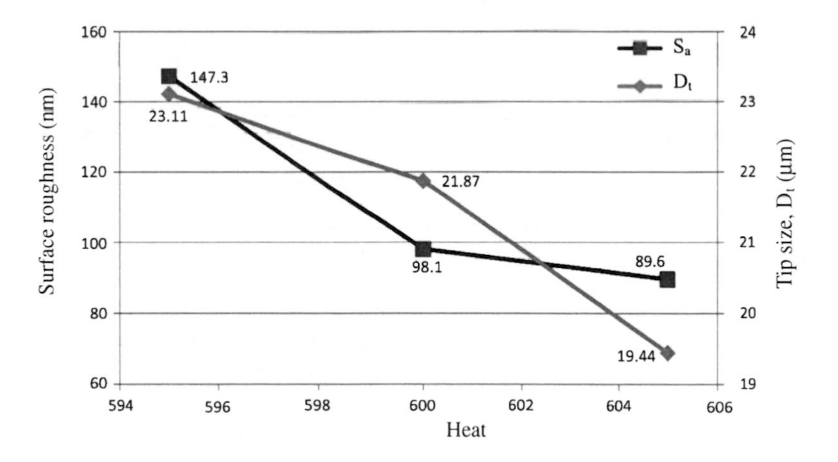

Fig. 7.4 The effect of heat on aperture size and average surface roughness. The heat is controlled by the level of electrical current supplied to the filament. The unit of heat is the milliamp. Useful changes in heat are 5 units or more to see an effect. By increasing the heat, both of the S_a and D_t decrease [14]

have a direct correlation i.e., where the tip size is increased, surface roughness also increases. This result is consistent with the results of Chap. 6 which states that smaller pipettes form a better seal.

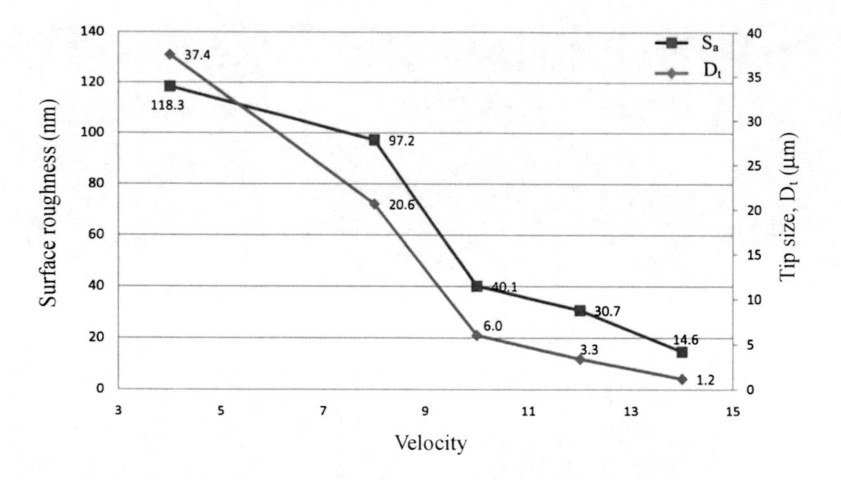

Fig. 7.5 The effect of velocity on aperture size and average surface roughness. This control measures the velocity of the glass carriage system as the glass softens. By increasing the velocity, both the tip size and the surface roughness decrease. The velocity has the most significant effect on the tip aperture size and the surface roughness. A small change in velocity value decreases S_a and D_t rapidly [14]

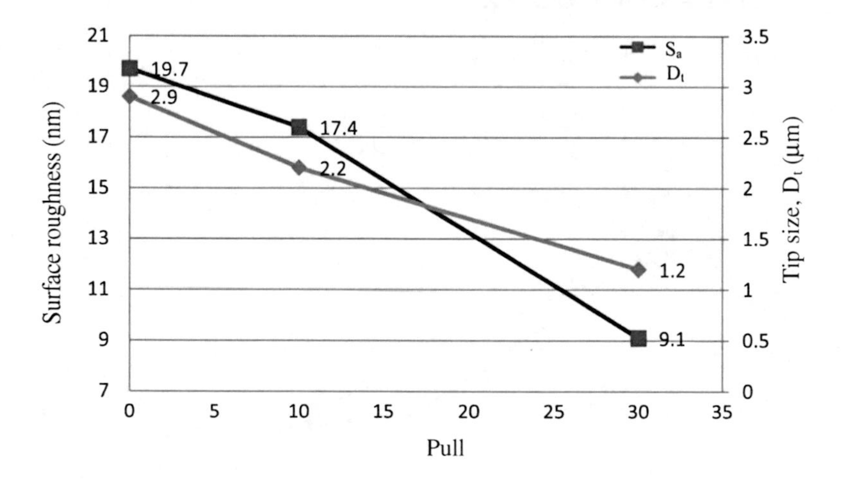

Fig. 7.6 The effect of pull on aperture size and average surface roughness. This parameter controls the force of the hard pull. The amount of the pull determines the current to the pull solenoid. Useful changes in pull strength are 10 units or more to see an effect. By increasing the pull, both of the S_a and D_t decreases [14]

7.3 Effect of Pulling Direction on Pipette Surface Properties

It has been believed that pulling pipettes with a heating and pulling process results in a smooth surface [15]. The surface properties of pipettes' inner walls were reported in Sect. 6.3.2 and the effects of pulling parameters are discussed in the previous section. Here in order to determine the effect of pulling direction on the surface texture of a pipette inner wall, autocorrelation of a roughness model for the pipette inner surface

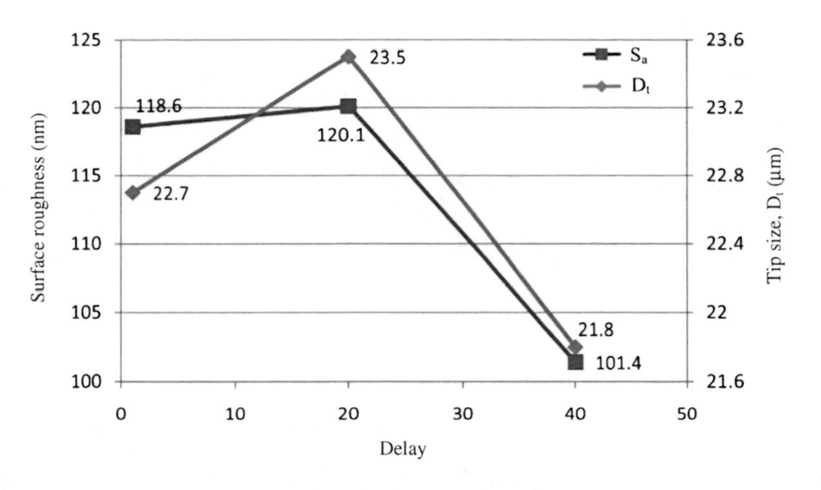

Fig. 7.7 The effect of delay on aperture size and average surface roughness. Delay is a cooling mode which controls the delay time between the time when the heat turns off and the time when the hard pull is activated. One unit of delay represents ½ ms. Delay is an effective means of controlling the pipette shank length which does not notably change the pipette aperture size [14]

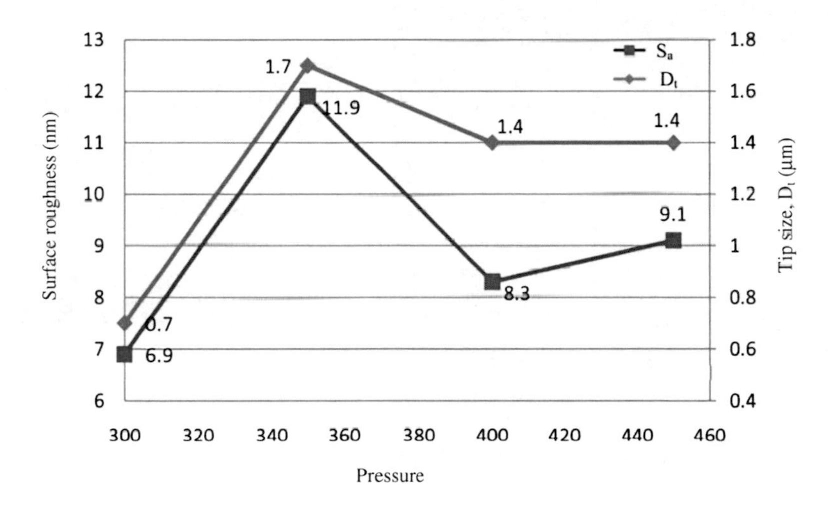

Fig. 7.8 The effect of pressure on aperture size and average surface roughness. This control sets the pressure generated by the air compressor during the active cooling phase of the pull cycle. The unit of pressure is psi. Changes of less than 10 units will not be noticeable. Pressure is another way of controlling the pipette shank length and does not significantly change the pipette aperture size [14]

is obtained. If the pulling direction is found to have an effect on the surface properties then it can be used as a tool for controlling the surface properties. Figures 7.10 and 7.11 show the pipette inner wall and its autocorrelation roughness model.

The autocorrelation plot suggests that the surface does not have any tendency towards orientation and is not affected by the pulling direction. This could be because micropipettes are fabricated by a heating and pulling process. Heating makes pipettes soften and no specific orientation can be achieved.

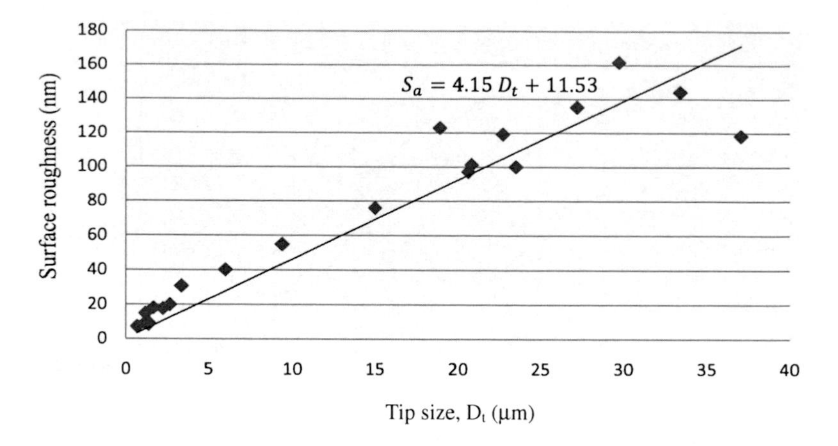

Fig. 7.9 Average surface roughness of pipette tip (S_a) versus aperture size (D_t). S_a is strongly dependent on D_t and has a direct correlation with it. A first-degree polynomial equation is fitted to the data [14]

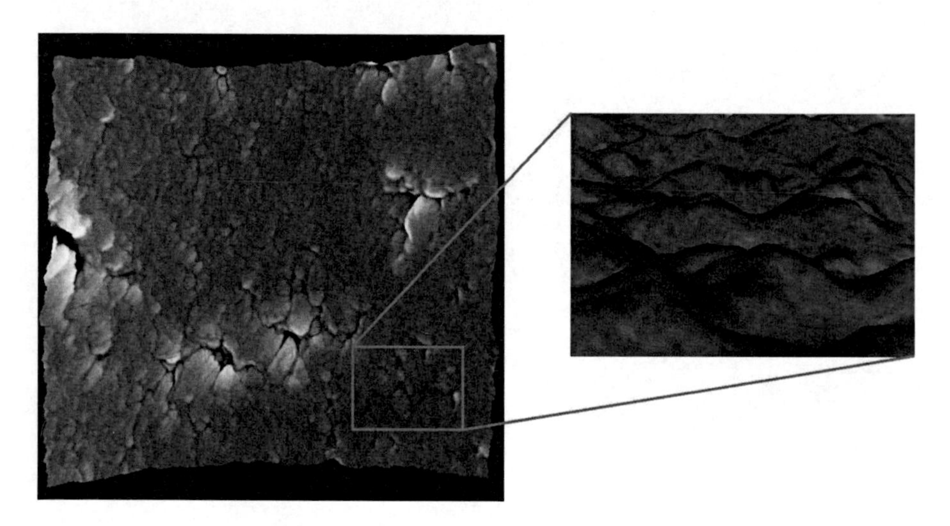

Fig. 7.10 Digital elevation model of the inner wall surface of the pipette shown in Fig. 6.9. As is shown in the inset the surface does not have a defined lay and consists of high frequency components

7.4 3D Reconstruction of a Pipette Using FIB/SEM Nanotomography

As was mentioned in Chap. 2, the geometry of the patching site is an important factor in seal formation. Ideally a smooth round shape is preferred since a patching site with sharp corners and irregular shapes is believed to increase leakage.

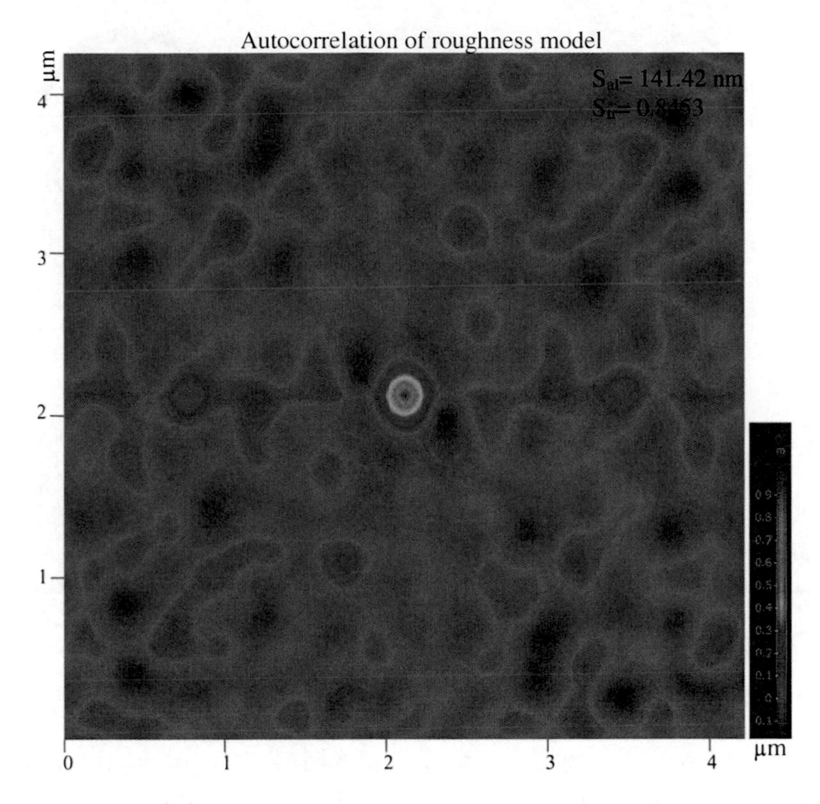

Fig. 7.11 Autocorrelation of roughness model of the pipette inner wall. The plot shows that the surface does not have any texture orientation. A high value for the Texture Aspect Ratio of the Surface (S_{tr}) indicates a uniform texture in all directions; i.e., no defined lay. A low value for the Autocorrelation Length (S_{al}) denotes that the surface is dominated by high frequency components [14]

Micropipettes are produced from glass tubes with a circular cross-section and it is assumed that the pipette tip is also circular. However, there has not been any report on the roundness of pipette tips in the literature. In gigaseal formation the cell membrane has contact with the last 100 microns of the pipette tip. Reconstruction of this area provides valuable information about the exact geometry of the contact area.

7.4.1 Effect of Omega Dot

A glass fibre is commonly fused along the inner bore of capillary tubing to facilitate the filling of micropipette tips with conducting solutions. This internal fibre is called Omega Dot [5]. The omega dot increases the capillary action and facilitates the filling of pipettes with the solution. Figure 7.12 shows the effect of omega dot on the shape of

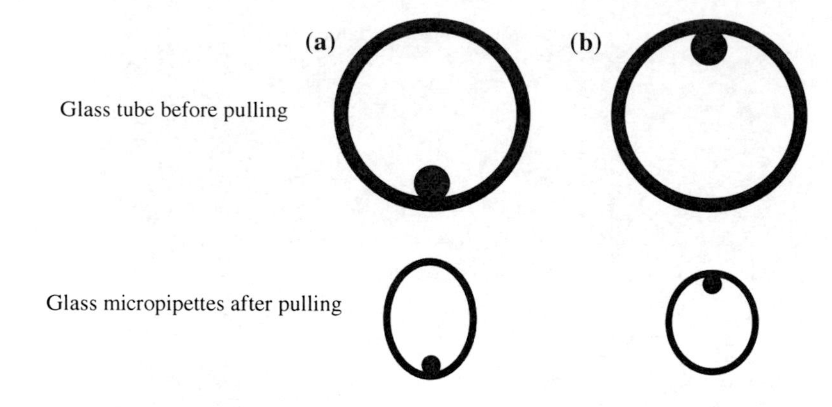

Fig. 7.12 Schematic of the cross-sections of micropipettes formed with omega dot either a) down toward the filament or b) upward away from it. Omega dot changes the circular cross-setion of the pipette to an elliptical shape. The effect is more significant with the omega dot down

the micropipette tip cross-section. Omega dot changes the circular cross-section of the pipette to an elliptical shape. The effect is more significant with the omega dot down.

7.4.2 3D Reconstruction of a Pipette

A FEI dual-beam Strata 235 focused ion beam (FIB) system was used as a nanotomography tool to obtain the 3D shape of a pipette tip. The process involves a cycle of milling a slice of the pipette using FIB, taking an SEM image of the new surface, and then milling and imaging again to produce a stack of SEM images. A micropipette with a tip size of 1.3 μm was placed facing the electron beam. Figure 7.13a shows the schematic of the pipette, electron beam and ion beam configuration. The angle between I-beam and E-beam was 52°. Therefore, the angle between the imaging plane and the sample was 38°, as seen in Fig. 7.13b. This information was used later for reconstruction. Each slice of the sample was milled off using Ga^+ ion beam at 30 kV and 100 pA for 90 s and dwell time of 1 μs with overlap parameters of 50 %. Sixty slices with a total thickness of 3 μm were removed and SEM images of the slices taken. The pixel size of the SEM images was 4.5 nm. Figure 7.14 shows an image of the 20th slice after milling and its internal edge.

In order to be able to reconstruct the tip in three-dimensions, first the following steps should be carried out:

- Image alignment
  A feature-based alignment method has been used [16]. A fixed feature which has not been milled during slicing and is not affected by the ion beam is the bottom left-hand side of the pipette in Fig. 7.14 which has been used for the alignment of the images.
- Edge detection
  The edge of the internal circle of the pipette was detected using the Canny algorithm [17].

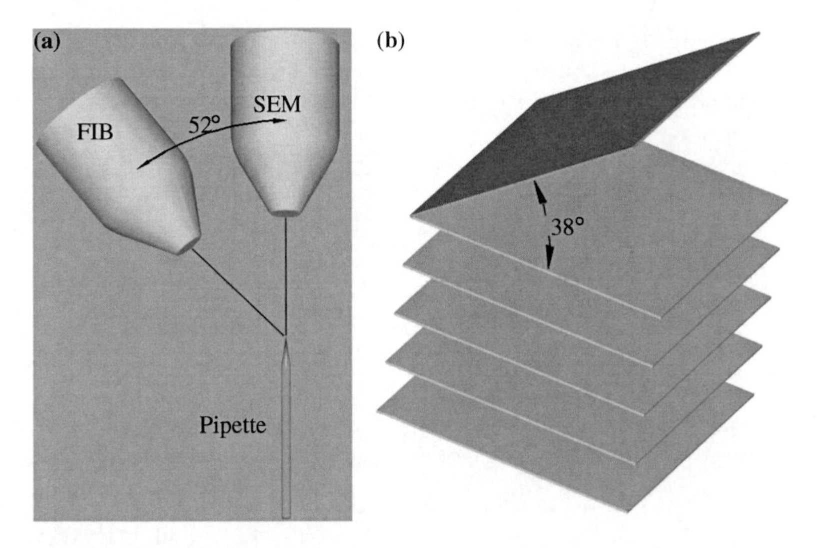

Fig. 7.13 FIB nanotomography of a glass micropipette. **a** schematic of pipette, E-beam and I-beam configuration, **b** a schematic of projected plane and the sample slices planes. The brown face shows the projected planes

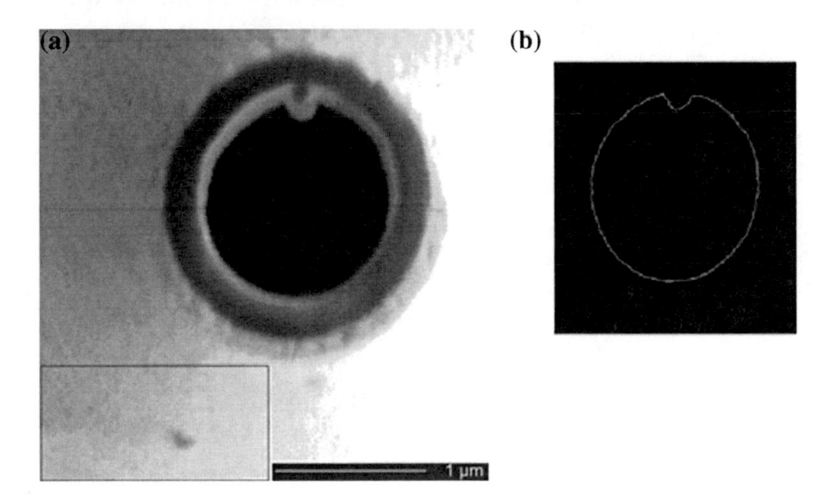

Fig. 7.14 Edge detection by using the Canny algorithm. **a** An SEM image of a pipette after milling, the rectangle represents the area which has not been milled during slicing and is not affected by the ion beam, **b** detecting the edge of the internal wall of the pipette using the Canny algorithm

The basic idea of this algorithm is to detect the zero-crossing of the second derivative of the smoothed images. It seeks out the zero-crossings of:

$$\partial^2(\mathrm{M} * \mathrm{I}) \Big/ \partial \mathrm{n}^2 = \partial \left(\left[\partial \mathrm{M} / \partial \mathrm{n} \right] * \mathrm{I} \right) / \partial \mathrm{n}$$

Where M and I are image matrix and unit matrix respectively and n is the direction of the gradient of the smoothed image. Edge detection was performed using the image processing tool box of MATLAB software for all of the slices.

- Back projection
  As far as the image plane has an angle of 38° with respect to the sample slices (Fig. 7.13) one can calculate the position of each point of the sample slice. Assuming that x (horizontal) and y (vertical) axes are in the image plane and z is the norm of the surface, then:

$$x_{ssp} = x$$
$$y_{ssp} = y/\cos 38^{\circ}$$

  where indices 'ssp' is for the sample slice position. For zssp, the thickness of the slices are 50 nm so the relative distance between the slices remains 50 nm (or 11 pixels) and the initial angle of the image plane and projected plane is 38°.

Figure 7.15 shows the 3D structure of the pipette tip reconstructed using MATLAB. The units of X, Y, Z axes are in pixels and each pixel is 4.5 nm. In order to examine the shape of the pipette tip, a perfect circle was fitted to each slice based on the least squared fitted circle method [18] and maximum deviation of the pipette shape from the circle was obtained. Figure 7.16 shows the first slice image and the fitted circle. The maximum deviation from the fitted circle is 43 nm for this slice. The average of maximum deviations of all slices was found to be 67 nm or 10 % in roundness error.

7.5 Tip Formation

Studies and observations carried out on the pipette's tip surface and geometry led into a new hypothesis for tip formation. In the literature two different mechanisms for micropipette tip separation have been discussed. One mechanism considers

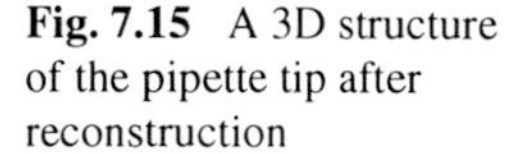

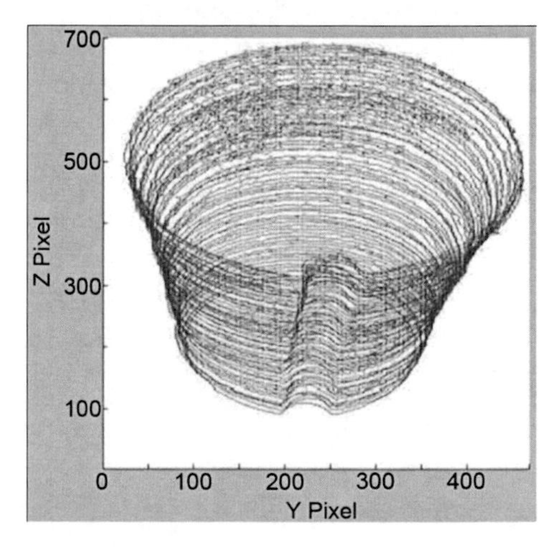

Fig. 7.15 A 3D structure of the pipette tip after reconstruction

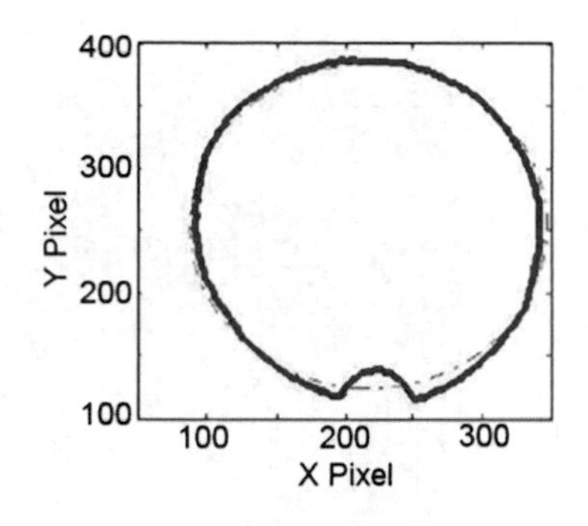

Fig. 7.16 An image of the first slice and a fitted circle. The fitted circle is shown in dashed line

that tips separate while they are in a fluid phase [5] and another mechanism considers that separation happens in a solid phase, by fracture [19]. Brown et al. have assumed that tip separation most probably happens while glass is in a fluid phase, which is partly because of the appearance of micropipette tips under high-resolution SEM. The tips are almost always formed at right angles to the long axis of the micropipette, and without major irregularities. This result may be expected if the two tips separate while still in the fluid phase and then harden shortly afterwards. In Brown et al.'s model, separation occurs when the thickness of the glass wall has become reduced to a point where it cannot be further attenuated at the prevailing viscosity [5]. On the other hand, in the model by Purves it is assumed that separation into two pipettes occurs by fracture when the stress exceeds the tensile strength of the glass [19]. If the tips are separated by fracture then one could expect inconsistent fracture orientation and irregular or jagged edges. In both models it was assumed that the temperature, and hence the viscosity, of the glass does not change during formation of the tips.

High-resolution SEM images together with the 3D reconstruction technique have made it possible to examine pipette tips precisely. Figure 7.17 shows SEM images of some large pipettes. Some common factors in all of them are cracks, inconsistent fracture orientation and irregular or jagged edges. These factors are clearly signs of fracture in the solid phase. This means that the tip is cooled down after the heat has turned off and fracture happens a result of the hard pulling of glass.

Figure 7.18 shows SEM images of some small pipettes. The pipettes have smooth and round tips. There is no sign of cracks. More than 20 pipettes were examined and it was found that similar to the results of Brown et al. [5], the tips are almost always formed at right angles to the long axis of the micropipette, and without major irregularities. This suggests that for small pipettes the tips are formed while the glass is still in the fluid phase.

By comparing images of pipettes in Figs. 7.17 and 7.18 it can be hypothesized that the mechanism of tip formation is dependent on the tip size. For large pipettes (tip size > 20 μm) tips are formed by fracture in a solid phase; while in the case of small pipettes (tip size < 2 μm) the tips are formed while the glass is still in a fluid phase.

For further investigation, tips of micropipettes were studied in more detail. If the tips are formed in a fluid phase then tip surface rearranges itself to minimize

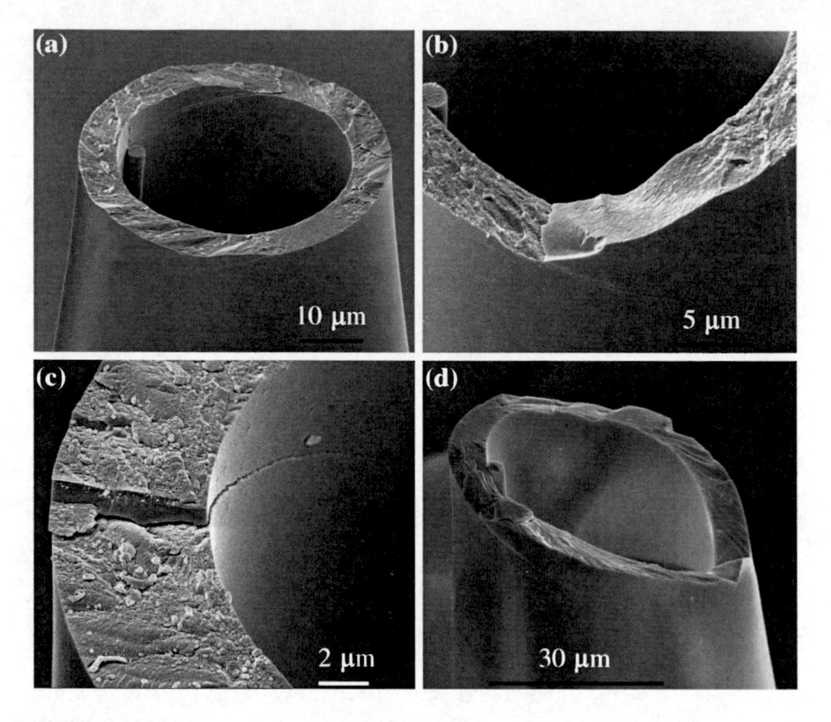

Fig. 7.17 SEM images of some large pipettes ($D_t > 20$ μm). Some distinctive features of large-sized tips are: cracks, irregular and rough tips, and inclined orientation of fracture

the free Gibbs energy; therefore, the tips should have a hemispherical shape. This has been schematically shown in Fig. 7.19. If the tips are formed by fracture, on the other hand, then irregular or jagged profiles are expected across the tip thickness.

In order to examine this hypothesis, tips of some large and small pipettes were 3D reconstructed using a SEM stereoscopic technique and profiles across their tip thickness were obtained. Figure 7.20 shows some tip profiles of a small pipette ($D_t = 2.7$ μm). As can be seen, the profiles are relatively smooth curves with a peak in the middle. This result may confirm that the tips are formed in a fluid phase.

Figure 7.21 shows some tip profiles of a large pipette ($D_t = 27.9$ μm). As can be seen, the profiles are irregular and jagged. This confirms the hypothesis that the tips of larger pipettes are formed in a solid phase.

The profiles are very close to the hemispherical shape for smaller pipettes, which confirms the hypothesis above. The results are also consistent with previous findings and could explain why larger pipettes have higher surface roughness. For tip sizes between 2 and 20 μm both kinds of feature were visible at the tip, and therefore this range can be considered as the transitional range going from solid fracture to fluid separation.

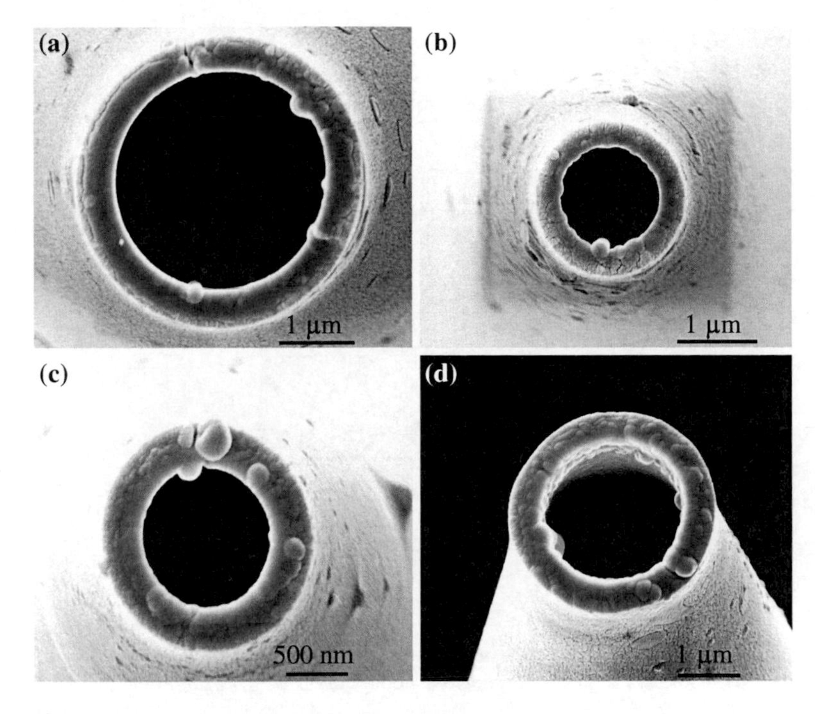

Fig. 7.18 SEM images of some small pipettes ($D_t < 2$ μm). Tips are flat and smooth and have dome shape

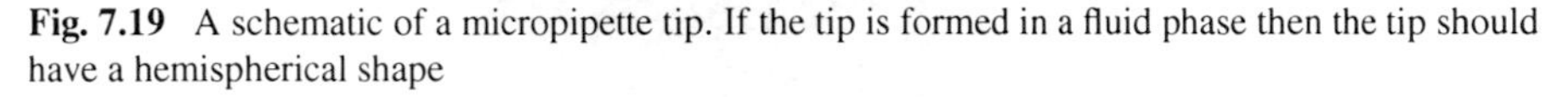

Fig. 7.19 A schematic of a micropipette tip. If the tip is formed in a fluid phase then the tip should have a hemispherical shape

7.6 Summary

In this chapter various aspects of glass micropipettes have been studied. Glass micropipettes are fabricated from glass tubes by a heating and pulling process. Heat, velocity, pull, delay and pressure are the controllable parameters of the fabrication process and the effects of these parameters on the tip size and surface roughness properties of pipettes were studied. It was found that there is a direct correlation between tip size and the surface roughness of the pipette, i.e., when the

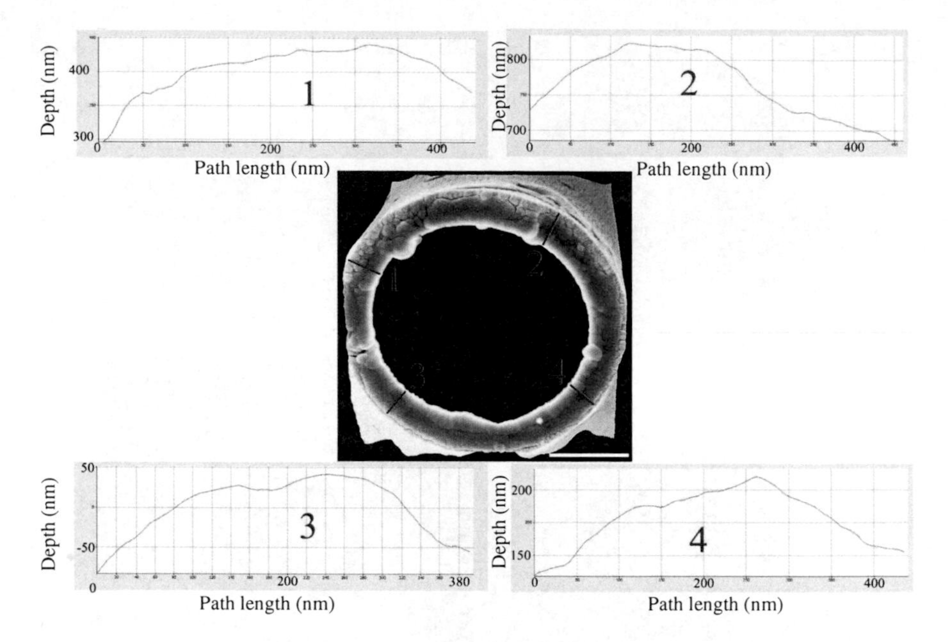

Fig. 7.20 Digital elevation model of a small pipette tip ($D_t = 2.7\ \mu m$) and four different profiles across the tip thickness. Profiles are relatively smooth curves with a peak in the middle. The bar represents 1 μm

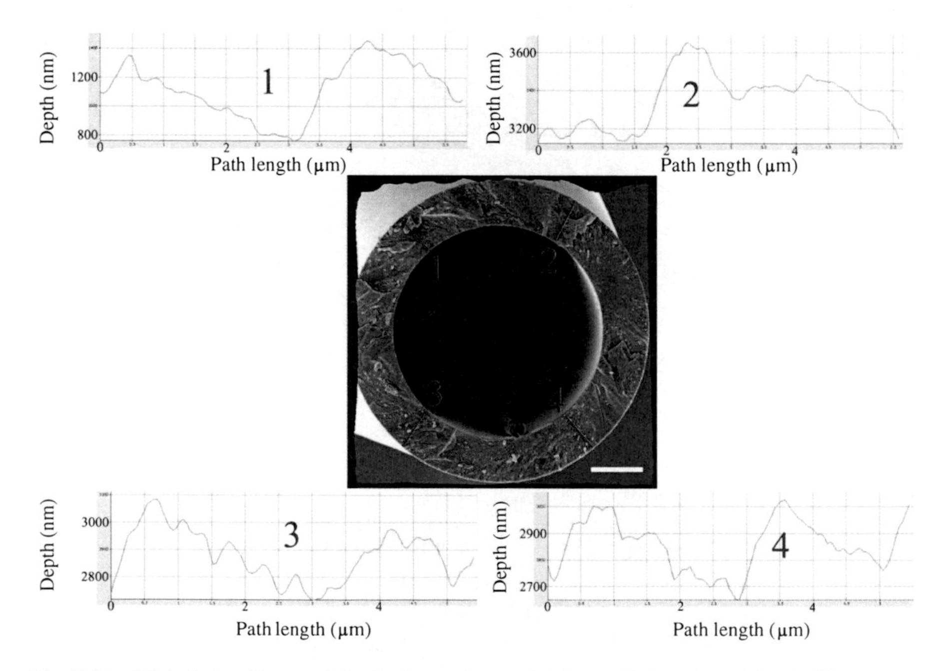

Fig. 7.21 Digital elevation model of a large pipette tip ($D_t = 33.1\ \mu m$) and four different profiles across the tip thickness. Profiles are irregular and jagged. The bar represents 10 μm

tip size is increased, surface roughness also increases. An autocorrelation plot of the inner wall surface of a pipette showed that the surface does not have any orientation tendency and is not affected by the pulling direction. The roundness of the pipette in the area of contact with a cell was also measured, using FIB nanotomography and image processing techniques. It was found that although the original glass tubes are circular in cross-section, the tips of micropipettes are not circular and have average maximum deviations of 67 nm (10 % in roundness error) for a micropipette with tip size of 1.3 μm. These studies have been valuable in fostering a better understanding of the mechanisms of tip formation in glass micropipettes. The results show that two different mechanisms are involved in glass micropipette tip formation. The mechanisms are dependent on the tip size of the pipettes to be formed. If the parameters are set to produce large pipettes then tips are formed by fracture in a solid phase; while if they are set to produce small pipettes, then the tips are formed in a liquid phase. The findings and results of this chapter explain sources of leakage in seal formation and can be useful in various applications of glass micropipettes where surface properties and sizes are important.

References

1. Bruckbauer A et al (2007) Nanopipette delivery of individual molecules to cellular compartments for single-molecule fluorescence tracking. J Biophys 93:3120–3131
2. Keith Martin RG, Klein RL, Quigley HA (2002) Gene delivery to the eye using adeno-associated viral vectors. Methods 28:267–275
3. Kimura Y, Yanagimachi R (1995) Intracytoplasmic sperm injection in the mouse. Biol Reprod 52:709–720
4. Yaul M, Bhatti R, Lawrence, S (2008) Evaluating the process of polishing borosilicate glass capillaries used for fabrication of in vitro fertilization (iVF) micro-pipettes. Biomed Microdevices 10:123–128
5. Brown KT, Flaming DG (1995) Advanced Micropipette Techniques for Cell Physiology. John Wiley & Sons, San Francisco
6. Neher E, Sakmann B (1976) Single-channel currents recorded from membrane of denervated frog muscle fibres. Nature 260:799–802
7. Huebner A et al (2008) Microdroplets: a sea of applications?. Lab Chip 8:1244–1254
8. Hong M H et al (2000) Scanning nanolithography using a material-filled nanopipette. Appl Phys lett 77:2604–2606
9. Ying L et al (2005) The scanned nanopipette: a new tool for high resolution bioimaging and controlled deposition of biomolecules. Phys Chem Chem Phys 7:2859–2866
10. Operational Manual P-97 Flaming/Brown Micropipette Puller Sutter instrument company. [Online] www.sutter.com
11. Flaming Dale G (1986) Method of forming an ultrafine micropipette. CA 4600424, 15-July 1986
12. Oesterle A (2009) Personal communication. Sutter instruments, 24 Jan 2009
13. Marinello F et al (2008) Critical factors in SEM 3D stereomicroscopy. Meas Sci Technol 19:1–12
14. Malboubi M, Gu Y, Jiang K (2011) Surface properties of glass micropipettes and their effect on biological studies. Nanoscale Res Lett 6:1–10
15. Lepple-Wienhues A et al (2003) Flip the tip: an automated, high quality, cost-effective patch clamp screen. Receptors Channels 9:13–17

16. Kubis AJ et al (2004) Focused-ion beam tomography. Metall Mater Trans A 35(7):1935–1943
17. Canny J (1986) A computational approach to edge detection. IEEE Trans Pattern Anal Mach Intell 8:679–714
18. Gander W, Golub GH, Strebel R (1994) Least-squares fitting of circles and ellipses. BIT Numer Math 34:558–578
19. Purves Robert D (1980 March) The mechanics of pulling a glass micropipette. Biophys J 29:523–530